Payment of the Full Rate of Special and Incentive Pays to Members of the Reserve Components

JAMES V. MARRONE, MICHAEL G. MATTOCK, BETH J. ASCH, HANNAH ACHESON-FIELD

Prepared for the Office of the Under Secretary of Defense for Personnel and Readiness
Approved for public release; distribution unlimited

RAND NATIONAL DEFENSE RESEARCH INSTITUTE

For more information on this publication, visit **www.rand.org/t/RRA669-1**.

About RAND

The RAND Corporation is a research organization that develops solutions to public policy challenges to help make communities throughout the world safer and more secure, healthier and more prosperous. RAND is nonprofit, nonpartisan, and committed to the public interest. To learn more about RAND, visit www.rand.org.

Research Integrity

Our mission to help improve policy and decisionmaking through research and analysis is enabled through our core values of quality and objectivity and our unwavering commitment to the highest level of integrity and ethical behavior. To help ensure our research and analysis are rigorous, objective, and nonpartisan, we subject our research publications to a robust and exacting quality-assurance process; avoid both the appearance and reality of financial and other conflicts of interest through staff training, project screening, and a policy of mandatory disclosure; and pursue transparency in our research engagements through our commitment to the open publication of our research findings and recommendations, disclosure of the source of funding of published research, and policies to ensure intellectual independence. For more information, visit www.rand.org/about/principles.

RAND's publications do not necessarily reflect the opinions of its research clients and sponsors.

Published by the RAND Corporation, Santa Monica, Calif.
© 2022 RAND Corporation
RAND® is a registered trademark.

Library of Congress Cataloging-in-Publication Data is available for this publication.

ISBN: 978-1-9774-0738-2

About This Report

In section 653 of the National Defense Authorization Act for Fiscal Year 2020, Congress called for a report on the extension to members of the reserve component (RC) of special and incentive (S&I) pays that are paid to active component (AC) members under sections 334, 334a, and 351 of Title 37 of the U.S. Code. In this report, we (1) estimate the cost of paying members of the RC, who perform service at the typical rate of four training periods (known as *drills*) of inactive duty per month (or active duty for less than a full month), the same full monthly rate of S&I pays, or *full rate*, that AC members receive for performing the S&I pay qualifying service for a full month; (2) estimate the number of RC members who would qualify for each form of S&I pay at the full rate; and (3) consider the feasibility and advisability of paying eligible members of the RC at the full S&I pay rate. (For the purposes of the study, *S&I pays* do not include recruiting, accession, affiliation, retention, or other bonuses and payments that are paid in a lump sum, installments at intervals other than a monthly rate, or a combination of both.)

The research reported here was completed in July 2021 and underwent security review with the sponsor and the Defense Office of Prepublication and Security Review before public release.

This research was sponsored by the Office of the Under Secretary of Defense for Personnel and Readiness and conducted within the Forces and Resources Policy Center of the RAND National Security Research Division (NSRD), which operates the National Defense Research Institute (NDRI), a federally funded research and development center sponsored by the Office of the Secretary of Defense, the Joint Staff, the Unified Combatant Commands, the Navy, the Marine Corps, the defense agencies, and the defense intelligence enterprise.

For more information on the RAND Forces and Resources Policy Center, see www.rand.org/nsrd/frp or contact the director (contact information is provided on the webpage).

Contents

Figures

Tables

Summary

Military service members are paid special and incentive (S&I) pays, such as hazardous duty incentive pay (HDIP) and aviation incentive pay (AvIP), on a prorated basis in proportion to the amount of basic pay they receive in a month. Typically, full-time active duty service members are assigned to the relevant duty for a full month, and so they receive the full monthly rate of the S&I pay. Typically, reserve component (RC) members serve for less than a full month and so receive a proportional amount of the pay. For example, if an RC member participates in four drill or training periods in a month (meaning two days of service) and performs qualifying duty for an S&I pay, they receive four-thirtieths of the full rate of the S&I pay.

Some observers argue that RC members should be paid at the same full monthly rate as typical active component (AC)[1] members, since RC members are required to undergo the same training as AC members, and the standards RC members must meet to establish proficiencies to receive certain S&I pays are the same as for AC members. For example, the eligibility requirement to receive HDIP for parachute duty is the same for AC and RC members: specifically, at least one jump in a three-month period. But AC members receive the full $150 per month for a static line parachute jump, while RC members receive a prorated amount of that $150. Others counter that RC members do not have the same readiness availability as AC members serving an entire month of duty and thus should not receive the full rate of S&I pay.

In response to this issue, Congress called for a report on the extension to members of the RC of S&I pays that are currently paid to AC members in section 653 of the National Defense Authorization Act for Fiscal Year 2020. Specifically, the section requires the Secretary of Defense to report the results of a study that provides

---

[1] *Active component* is abbreviated AC in this report, but it also known as the "regular" component.

1. an estimate of the yearly cost of paying members of the RC risk pay and flight pay under sections 334, 334a, and 351 of Title 37 of the U.S. Code at the same rate as members of the AC

2. an estimate of the number of RC members who would qualify or would potentially qualify for HDIP, according to current professions and duties and broken out by hazardous duty categories.

These reporting requirements are inputs to a required assessment of the feasibility and advisability of paying eligible members of the RC any S&I pay that is currently paid to members of the AC.

Approach

Our approach was to use a combination of empirical, simulation, and theoretical techniques. For the estimate of cost, the first item required by Congress, our empirical approach was to gather data on when these S&I pays were being paid to RC members and compute how much these members would have been paid if they were paid the full monthly rate regardless of qualifying service. We call this the *full monthly rate* policy. In terms of data, we accessed from the Defense Manpower Data Center the pay records and duty status information for all RC members who served from January 2018 to March 2020. These dates were chosen so that we could track how special pays were disbursed from the beginning of the Hazard Pay program in 2018 until the beginning of closures in response to the coronavirus disease 2019 (COVID-19) pandemic in March 2020. We also received information on annual flying hours for RC members from the individual RCs. Information on flying hours is required because eligibility for certain S&I pays depends on flying hours, and the requirement differs for AC and RC personnel.

Our empirical approach was complicated by the fact that available data did not precisely identify the month when a duty eligible for S&I pay was performed. Pay files may show payments for multiple months of service being received in a single month. Files showing duty status do not include inactive duty training (IDT) and thus do not show all months when an individual is performing duty eligible for an S&I pay. In response to these data challenges, we developed three approaches to approximate the number of months an individual would be eligible for S&I pays and used these approaches to calculate a range of estimates for the yearly cost of paying RC members the full monthly rate of S&I pays.

For the second item required by Congress, we calculated potential eligibility in two ways: We (1) counted the number of RC members who earn each pay in an average month and (2) identified "eligible" occupations using the fraction of those earning each pay who were assigned to a duty occupation and then calculated how many RC members are assigned to those eligible occupations each month.

The empirical approach we used implicitly assumes that individuals would carry on as they had before if a full-rate policy were introduced. However, individuals could change their participation behavior in the RC in response to such a policy change, potentially affecting RC readiness and efficiency. To explore this issue, we supplemented the empirical analysis in two ways. First, we conducted simulations using the RAND Corporation's dynamic retention model (DRM) for U.S. Air Force pilots to assess the effect of a change in policy on AC retention and RC participation. Air Force pilots are a good example because they qualify for AvIP, one of the S&I pays of interest to Congress that accounts for a large share of S&I pay costs among the group of S&I pays under consideration by Congress. Second, we extended a theoretical model by Shishko and Rostker, 1976, to consider the effect of a full-rate policy on the intensity of participation by current RC members (number of drills or days served in a month) and the attractiveness of the RC to individuals from a recruiting standpoint.

Findings

Our first finding is that the yearly cost increase associated with paying members of the RC risk pay and flight pay under sections 334, 334a, and 351 of Title 37 of the U.S. Code at the full monthly rate would range from $46.3 million to $88.5 million annually, or 100 percent to 194 percent over the baseline of $45.7 million annually.

Our second finding is that RC members serve more periods per month than the stereotypical one weekend per month and two weeks in the summer. We refer to *periods* of service because RC members serve in units of days when on an active status and in four-hour drill periods when on an inactive status. We estimate that RC members serve an average of 6.1 periods across all duty statuses in a month, and those who earn certain pays (such as AvIP) serve even more. If RC members served the stereotypical amount each month, they would serve an average of 5.2 periods per month. Because RC members serve more than the required minimum number of periods each month, their S&I payments each month are higher than what we would expect from a stereotypical reservist, reflecting the greater than minimum intensity of service. The implication is that the cost increase associated with a full-rate S&I pay policy is less than we would have expected because RC members are already paid closer to the full monthly rate than would be expected from just four IDT drills per month and 14 days of annual training per year.

Third, we find that the number of RC members who would qualify or would potentially qualify for either HDIP or AvIP in a given month ranges from 17,796 to 379,148. The lower estimate reflects the number of RC members who earned these pays over our data period, while the upper estimate reflects the number of RC members in occupations with a large share of personnel who earn the pays. The difference in the two estimates shows that although a duty occupation may require that a member be trained and qualified to perform a hazardous duty, most RC members in that occupation do not actually meet the minimum performance requirement necessary to earn the pay.

Our exploratory analysis using simulation and theory provides insights into how RC member participation behavior might change if RC S&I pays are paid at the full rate. We find that the full-rate policy could reduce incentives to participate in the RC for more than the minimum required training periods: i.e., individuals might reduce the intensity of participation in each month (fewer drills or days per month). As noted earlier, we find that RC members perform more service than the minimum each month; the current policy of prorating S&I pay gives these RC members an incentive to increase the intensity of participation because the S&I pay is proportional to the amount of participation, up to the maximum S&I pay cap. The full-rate policy removes this relationship once the individual has qualified for the S&I pay and served the minimum required amount of service. Members who serve less than this amount under the current policy would accumulate less experience in their occupations. Thus, the full-rate policy would adversely affect readiness. Furthermore, the full monthly rate policy would be inefficient because it would increase costs while potentially reducing participation—i.e., paying more for less.

We also find that the full-rate policy would increase the propensity for individuals to join those RC occupations in which the full rate of S&I pay is being paid because of the increase in monthly earnings once members qualify for the S&I pay. This increase in the number who participate could help offset the drop in the intensity of participation of those who join to yield the same total number of person-days or person-drills. However, an increase in the number of personnel would likely further increase cost, beyond what we would predict when we assume individual behavior is fixed. Thus, the "maximum" 194 percent increase in cost projected could be too conservative.

Finally, our simulations using RAND's DRM indicate that AC retention would fall in the occupations that offer a full rate to RC members. In short, the full-rate policy in the RC would pull people out of the AC and into the RC. Thus, another adverse readiness effect is the potential for a drop in AC retention.

Conclusions

The congressional requirement that motivated this study included a mandate to assess the feasibility and advisability of paying eligible RC members full-rate S&I pays. To assess feasibility, we put our estimates of the costs and numbers of personnel who might be affected by a full-rate policy into a larger context in terms of the overall size and budgetary cost of RC personnel. We find that although the cost increase would represent a substantial increase in the S&I pay budget, it would be small relative to the overall RC personnel budget: less than 0.4 percent for fiscal year 2021. Furthermore, as a share of the RC as of March 2020 (which numbers more than 800,000), the number of RC personnel who earn HDIP is less than 1 percent for any given pay, whereas the number who could potentially earn HDIP is, at most, 20 percent of the RC force size for any given pay. We note that our assessment does not consider the administrative costs of implementing a full-rate

RC policy. To assess advisability, we considered the implications of a full-rate policy for readiness and efficiency, as discussed earlier.

We conclude that paying RC members the full rate of S&I pays could be feasible in terms of cost, given that cost (excluding implementation costs) would be quite small relative to the size of the RC personnel budget. However, paying RC members the full rate of S&I pays may not be advisable because it would be inefficient and could adversely affect readiness. In particular, paying RC members the full rate of S&I pays would not be efficient if the goal is to provide an incentive to participate for more than the minimum number of drills or days in a month. In fact, it would have the opposite effect: As long as individuals positively value leisure, increasing their wealth by the amount of the full rate of S&I pay each month would give individuals an incentive to reduce the time they devote to RC participation.

Acknowledgments

We thank many individuals who contributed to and supported this project. First, we thank Lernes Hebert, Deputy Assistant Secretary of Defense for Military Personnel Policy. We would also like to thank our project monitors, Bill Dougherty and COL Charlette Woodard, both in Military Compensation Policy in the Office of the Under Secretary of Defense for Personnel and Readiness (OUSD[P&R]), as well as Jeri Busch, Director of Military Compensation Policy in the OUSD(P&R), for sharing insight and guidance for this report. We thank our RAND colleagues John Winkler and Molly McIntosh for their help and support with this report. We thank Lisa Harrington and Tom Bush for serving as technical peer reviewers.

Abbreviations

AC	active component
ADT	active duty for training
AT	annual training
AvB	Aviation Bonus
AvIP	aviation incentive pay
DMDC	Defense Manpower Data Center
DoD	U.S. Department of Defense
DoDI	Department of Defense Instruction
DRM	dynamic retention model
FY	fiscal year
HALO	high altitude, low opening
HDIP	hazardous duty incentive pay
HFP	hostile fire pay
HzP	hazard pay
IADT	initial active duty for training
IDP	imminent danger pay
IDT	inactive duty training
NDAA	National Defense Authorization Act
OFD	operational flying duty
PFD	proficiency flying duty

QRMC	Quadrennial Review of Military Compensation
RC	reserve component
RCC	reserve component category
S&I	special and incentive
TRC	training and retirement category
VBSS	visit, board, search, and seizure
WMDCS	weapons of mass destruction civil support
YAS	years of aviation service

CHAPTER ONE

Introduction

Section 653 of the National Defense Authorization Act (NDAA) for Fiscal Year (FY) 2020 directed the U.S. Department of Defense (DoD) to conduct a study and the Secretary of Defense to submit a report to the congressional defense committees on the feasibility and advisability of extending to eligible members of the reserve component (RC) any special and incentive (S&I) pays that are paid to active component (AC) members and that are not currently paid to members of the RC. The RC comprises seven components, with the purpose of providing "trained units and qualified persons available for active duty in the armed forces, in times of war or national emergency" (10 U.S.C. § 10102).[1] The Selected Reserves are a major part of the RC and consist of personnel who are generally required to perform one weekend of training each month—*inactive duty training* (IDT)—and two weeks of training each year—*annual training* (AT). If not called to active duty, Selected Reservists serve in the military part time, and many also hold full-time civilian jobs. Selected Reservists who are assigned to train with particular units and are not otherwise assigned to an AC billet or who serve as unit support are often called *drilling reservists*.

RC members are paid according to the same pay table as AC members, based on rank and years of service. However, they are compensated using different units of time. During AT, reservists receive one day of basic pay for each day of duty, just like AC members. But they receive one day of basic pay for each drill period during IDT, and there are two drill periods per week-

---

[1] The seven components are the Army National Guard, the Army Reserve, the Marine Corps Reserve, the Air National Guard, the Air Force Reserve, the Air National Guard, and the Coast Guard Reserve. A description of the RC, including pay and benefits, is given in Kapp and Torreon, 2020.

end day (morning and afternoon), or four drill periods per weekend. Consequently, RC members receive four days of basic pay for a two-day drill weekend. RC personnel may also be eligible for S&I pays if they perform certain types of hazardous or arduous duties, serve in certain assignments, or possess certain skills. For RC members, these monthly payments are generally prorated according to the number of days or drill periods performed using the proportionate rate of compensation received under 37 U.S.C. § 206, which uses a 30-day month, commonly referred to as the *one-thirtieth rule*. For example, an eligible reservist performing the standard two weeks of AT would receive fourteen-thirtieths of the monthly S&I pay rate. Eligible reservists performing standard four-drill IDT (weekend drill) would receive four-thirtieths of the monthly S&I rate.

Some observers argue that prorating S&I pay for drilling reservists is unfair to RC members and that RC members should be paid the full rate regardless of how many periods they serve in a month (Levinsky, 2020; Titus, 2020).[2] Their argument is that RC members are required to undergo the same training as AC members, and the standards RC members must meet to establish proficiencies to receive certain S&I pays are the same as for AC members. For example, the eligibility requirement to receive hazardous duty incentive pay (HDIP) for parachute duty is the same for AC and RC members (specifically, at least one jump in a three-month period). But AC members receive the full $150 per month for a static line parachute jump, whereas RC members receive a prorated amount of that $150.

A counterargument would be that AC members also receive prorated pays. They receive the full amount only when their actual service during the month is 30 days (or 28 days in February). Moreover, an RC member

---

[2] Chapter Two provides more details on the different S&I pays under 37 U.S.C. § 334, 334a, and 351 and whether and how eligibility criteria differ for AC versus RC personnel. For example, in some cases, such as with aviation incentive pay, eligibility depends on flying hours, and the number required for RC eligibility is half that for AC eligibility. For other cases, such as with hazardous duty incentive pay for parachute duty, AC and RC eligibility criteria are the same, but RC pays are prorated by the number of imminent danger pay drills or active training days, whereas AC pays are not. In addition, for other pays, such as hostile fire pay, there is no difference in eligibility criteria or the pay amounts for AC and RC members. Also, as discussed in Chapter Two, AC members receive prorated S&I pay in certain circumstances and for certain S&I pays. For example, AC members who separate midmonth will receive prorated S&I pay.

receives twice the daily compensation as an active duty member during IDT and therefore receives twice the amount of S&I pay for those qualifying days. Per this counterargument, S&I pays should be linked to days of qualifying duty rather than to the number of times a member receives basic pay in a month.

Section 653 of the FY 2020 NDAA directed the Secretary of Defense to investigate the feasibility and advisability of paying RC members at the full monthly rates of S&I pays, focusing specifically on hazard pay (HzP) and aviation incentive pay (AvIP; otherwise known as *flight pay*) under 37 U.S.C. § 334, 334a, and 351. HzP includes an array of risk-related pays, including hostile fire pay (HFP), imminent danger pay (IDP), and several categories of HDIP. Specifically, Congress requested

1. an estimate of the yearly cost of paying RC members HzP and AvIP at the same rate as AC members, regardless of the number of periods of instruction or appropriate duty participated in, so long as there is at least one such period of instruction or appropriate duty in the month
2. a statement of the number of RC members who qualify or potentially qualify for HDIP according to current professions or required duties, broken out by hazardous duty categories set forth in 37 U.S.C. § 351.

The purpose of this report is to provide input to the Office of the Secretary of Defense for its report to Congress. Specifically, we estimate the cost of paying RC members who perform IDT or active duty service for less than a full month the same monthly rate of S&I pays as AC members receive for a full month. In addition, we estimate the number of RC members who would qualify for each form of S&I pay at the full rate and consider the feasibility and advisability of paying eligible members of the RC at the full S&I pay rate, which is not currently payable to them. For the purposes of this report, *S&I pays* do not include recruiting, accession, affiliation, retention, or other bonuses and payments that are paid in a lump sum, installments at other than a monthly rate, or a combination of both.

To conduct our analysis, we draw from two data sources: (1) Defense Manpower Data Center (DMDC) data on RC pay and statuses for duty performed by RC members and (2) service-provided data on the qualification of RC members for AvIP and HDIP for flying duty. Because of data limitations, we can only approximate the effect of changing to a full-rate policy, because neither the DMDC data on RC pay nor the data on duty statuses provide complete data on what months an individual served IDT or AT. Because the month an individual served a particular duty has no bearing on what they get paid under the current proration policy, this information is not recorded in the pay file, and the duty statuses data do not record inactive duty statuses that might entitle an individual to S&I pay. Should Congress adopt a full-rate policy, the amount of pay would depend on when personnel perform service; members whose service is spread across two months would get two months of pay, while those whose service was conducted within one month would get only one month of S&I pay. We describe the data issues and our approach later in this report.

This report is organized as follows. Chapter Two provides relevant background material on the setting of S&I pay for RC members and the rules regarding the prorating of pay for each of the S&I pays defined under 37 U.S.C. § 334, 334a, and 351. In Chapter Three, we describe the data sources and how we merge the data to create analysis files that are structured to address the questions posed by Congress. We also describe our research approach. Chapter Four presents our estimates of the yearly cost of paying RC members the full rate of HDIP and AvIP, and Chapter Five presents tabulations of the number of RC members who potentially qualify for HDIP. In Chapter Six, we discuss the broader issue of equity of pay between the AC and RC, because the payment of S&I pays is not the only area in which AC and RC compensation differs. This chapter also provides some exploratory estimates of the effects on AC retention and RC participation of setting S&I pay at the full rate for RC members. We provide concluding thoughts in Chapter Seven with respect to the advisability and feasibility of paying eligible members of the RC at the full S&I pay rate.

CHAPTER TWO

Background

In this chapter, we describe the S&I pay included in the language of section 653 of the FY 2020 NDAA: specifically, in 37 U.S.C. § 334 and 334a (which cover AvIP) and 37 U.S.C. § 351 (which covers HzP). We provide an overview of each pay, including a description of each pay's eligibility requirements and authorized amounts for AC and RC personnel.[1] In this report, we focus on RC members who are performing IDT and AT duties. During IDT, RC members typically participate in at least four drills, one weekend per month. During AT, RC members participate in training that lasts for at least two weeks. In Appendix A, we provide more detail about these duty statuses, how reservists are compensated for their time generally, and how this compensation applies to S&I pay.

A summary of each pay is provided in Table 2.1. The table shows that eligibility for HzP and AvIP can depend on component and duty status. Our analysis focuses on those pays that RC members can earn when training and for which eligibility differs from AC members. We exclude pays for which RC members are already paid the same rate as AC members and face the same eligibility criteria, as well as pays that RC members cannot earn when training. In this chapter, we explain when this eligibility occurs and provide a list of the pays that are included in our analysis. In addition, we describe how the RC is structured, noting which RC members may be eligible for these special pays and the conditions under which they could earn the pays. Finally, we describe duty statuses in more detail.

---

[1] Two underlying factors affect the amount of pay that an RC member would receive in S&I pay: (1) the pay authority under which they serve and (2) their duty status. We discuss these two factors in Appendix A.

TABLE 2.1

S&I Pay Eligibility and Authorized Benefit for AC and RC Members

Pay	AC Eligibility and Training Requirement	AC Authorized Benefit[a]	RC Eligibility and Training Requirement for AT and IDT	RC IDT Authorized Benefit[b]	RC AT Authorized Benefit	Difference Between AC and RC	DoDI, DoDI Section
AvIP[c]	Must be an officer with an aeronautical rating that qualifies for OFD or PFD; can also qualify based on previous OFD and PFD service; must fly for 4 hours per month or 24 hours in 6 months[d]	Maximum is $1,000 but varies according to service and YAS (see Table A.3)	Same as AC, except monthly flying requirements are 2 hours per month or 12 hours in 6 months[d]	1/30th of the authorized monthly amount for each period of IDT[e]	Receive 1/30th of the authorized monthly amount for each day of AT	RC monthly flying requirement is half that of AC; RC is paid at 1/30th the monthly amount for each period of IDT	DoDI 7730.67, 3.2 and 3.3
HFP[f]	Must perform duty in an area that was certified to have hostile fire or be exposed to a hostile fire event during a given month	$225 (full monthly amount is paid even if exposure is for a partial month)	Same as AC	Same as AC	Same as AC	None	DoDI 1340.09, 3.1.c and 3.2
IDP[f]	Must be permanently assigned or perform duty in a designated imminent danger area	$225, IDP is prorated at 1/30th per day for every day they perform duty in the designated area	Same as AC	RC members are not eligible because they are prohibited from performing IDT in an imminent danger area	Same as AC	RC members are not eligible on an inactive duty status	DoDI 1340.09, 3.1.c and 3.3; DoDI 1215.06, Encl. 3, 2.1.d

Table 2.1—Continued

Pay	AC Eligibility and Training Requirement	AC Authorized Benefit[a]	RC Eligibility and Training Requirement for AT and IDT	RC IDT Authorized Benefit[b]	RC AT Authorized Benefit	Difference Between AC and RC	DoDI, DoDI Section
HDIP, flying duty	Must participate in 4 hours of aerial flight each month[b]	$250 for aircrew members; $150 for non-aircrew members	Must participate in 2 hours of aerial flight each month[b]	Monthly amount is paid in proportion to their monthly compensation[g]	Receive 1/30th of the authorized monthly amount for each day of annual training	AC must participate in 4 hours of aerial flight each month, whereas RC must participate in 2 hours per month; RC amount is prorated	DoDI 1340.09, 3.4.c
HDIP, parachute duty	Must perform duty involving parachute jumping and jump at least once in a 3-month period or twice in a 6-month period	$150 per month for static line; $225 per month for freefall	Same as AC	Monthly amount is paid in proportion to their monthly compensation[g]	Receive 1/30th of the authorized monthly amount for each day of annual training	RC amount is prorated based on the number of drills for IDT and days for AT	DoDI 1340.09, 3.4.d
HDIP, demolition duty	Must perform demolition using explosive objects, obstacles, or other explosives at least once per month	$150	Same as AC	Monthly amount is paid in proportion to their monthly compensation[g]	Receive 1/30th of the authorized monthly amount for each day of annual training	RC amount is prorated based on the number of drills for IDT and days for AT	DoDI 1340.09, 3.4.e
HDIP, experimental stress duty	Must perform as an experimental subject experiencing human acceleration or deceleration, thermal stress, a low-pressure chamber, or a high-pressure chamber	$150	Same as AC	Monthly amount is paid in proportion to their monthly compensation[g]	Receive 1/30th of the authorized monthly amount for each day of annual training	RC amount is prorated based on the number of drills for IDT and days for AT	DoDI 1340.09, 3.4.f

Table 2.1—Continued

Pay	AC Eligibility and Training Requirement	AC Authorized Benefit[a]	RC Eligibility and Training Requirement for AT and IDT	RC IDT Authorized Benefit[b]	RC AT Authorized Benefit	Difference Between AC and RC	DoDI, DoDI Section
HDIP, flight deck hazardous duty[h]	Must perform 4 days of flight operations per month on the flight deck of eligible air-capable ships	$150	Same as AC	Monthly amount is paid in proportion to their monthly compensation[g]	Receive 1/30th of the authorized monthly amount for each day of annual training	RC amount is prorated based on the number of drills for IDT and days for AT	DoDI 1340.09, 3.4.g
HDIP, duty involving exposure to highly toxic pesticides	Must perform duty that requires frequent exposure to toxic pesticides in an entomology, pest control, pest management, or preventative medicine role; must be assigned to the duty for 30 or more consecutive days	$150	Same as AC[i]	Monthly amount is paid in proportion to their monthly compensation[g]	Receive 1/30th of the authorized monthly amount for each day of annual training	RC amount is prorated based on the number of drills for IDT and days for AT	DoDI 1340.09, 3.4.h
HDIP, laboratory duty utilizing live dangerous viruses or bacteria	Must perform basic or applied research with live, dangerous viruses or bacteria that have a high potential for mortality and no prophylactic vaccine; must be assigned to the duty for 30 or more consecutive days	$150	Same as AC[i]	Monthly amount is paid in proportion to their monthly compensation[g]	Receive 1/30th of the authorized monthly amount for each day of annual training	RC amount is prorated based on the number of drills for IDT and days for AT	DoDI 1340.09, 3.4.i

Table 2.1—Continued

Pay	AC Eligibility and Training Requirement	AC Authorized Benefit[a]	RC Eligibility and Training Requirement for AT and IDT	RC IDT Authorized Benefit[b]	RC AT Authorized Benefit	Difference Between AC and RC	DoDI, DoDI Section
HDIP, duty involving toxic fuels and propellants	Must use toxic fuels and propellants for servicing and testing aircraft or missiles as part of the primary duty	$150	Same as AC[i]	Monthly amount is paid in proportion to their monthly compensation[g]	Receive 1/30th of the authorized monthly amount for each day of annual training	RC amount is prorated based on the number of drills for IDT and days for AT	DoDI 1340.09, 3.4.j
HDIP, duty involving handling chemical munitions	Must handle chemical munitions or surety material as part of the primary duty	$150	Same as AC[i]	Monthly amount is paid in proportion to their monthly compensation[g]	Receive 1/30th of the authorized monthly amount for each day of annual training	RC amount is prorated based on the number of drills for IDT and days for AT	DoDI 1340.09, 3.4.k
HDIP, maritime visit, board, search, and seizure (VBSS) duty	Must be assigned to a VBSS billet for a full month, regularly participate in VBSS operations, participate in at least 3 boarding missions per month, and be properly trained in VBSS operations	$150	Must serve in a VBSS billet for a full month; thus, RC members serving IDT and AT are not eligible	N/A	N/A	RC members serving IDT or AT are not eligible	DoDI 1340.09, 3.4.l

Table 2.1—Continued

Pay	AC Eligibility and Training Requirement	AC Authorized Benefit[a]	RC Eligibility and Training Requirement for AT and IDT	RC IDT Authorized Benefit[b]	RC AT Authorized Benefit	Difference Between AC and RC	DoDI, DoDI Section
HDIP, polar region flight operations duty	Must use ski-equipped aircraft in Antarctica or the Arctic ice pack and participate in at least one related flight or service related cargo	$150	Same as AC[i]	Monthly amount is paid in proportion to their monthly compensation[g]	Receive 1/30th of the authorized monthly amount for each day of annual training	RC amount is prorated based on the number of drills for IDT and days for AT	DoDI 1340.09, 3.4.m
HDIP, weapons of mass destruction civil support (WMDCS) team	Must be a member of a WMDCS team, be fully qualified for WMDCS team operations, and serve in an AC tour of at least 139 days	$150	RC members serving IDT and AT are not eligible	N/A	N/A	RC members serving IDT and AT are not eligible	DoDI 1340.09, 3.4.n
HDIP, diving duty	Must perform frequent and regular dives as part of the primary duty; must be entitled to basic pay	$240[j]	RC members on AT eligibility are same as AC; RC members on IDT are not eligible[k]	N/A	Receive 1/30th of the authorized monthly amount for each day of annual training	RC members serving IDT are not eligible; no difference otherwise for AT	DoDI 1340.09, 3.4.o

Table 2.1—Continued

Pay	AC Eligibility and Training Requirement	AC Authorized Benefit[a]	RC Eligibility and Training Requirement for AT and IDT	RC IDT Authorized Benefit[b]	RC AT Authorized Benefit	Difference Between AC and RC	DoDI, DoDI Section

SOURCES: DoDI 1215.06, 2015; DoDI 7730.67, 2016; DoDI 1340.09, 2018; Office of the Under Secretary of Defense (Comptroller), 2020.

NOTE: DoDI = Department of Defense Instruction; N/A = not applicable; OFD = operational flying duty; PFD = proficiency flying duty; YAS = years of aviation service.

[a] This table assumes that all AC members serve the full month. In reality, if an AC member serves a partial month, they receive a prorated amount of basic pay and would receive a prorated amount of S&I pay. Appendix A describes this in more detail.

[b] S&I pay is prorated according to the amount of compensation an RC member receives each month, which corresponds to the number of drills they perform each month for IDT. Therefore, our table applies to RC members performing any number of drills each month. Appendix A describes this in more detail.

[c] Service members who receive AvIP may not receive an HDIP for the same skill and period of service (i.e., HDIP for flying duty).

[d] Flying hours performed on an active status are typically considered separate from hours performed on an inactive duty status. However, if a service member serves in both an active and inactive status in the same month, the flying requirements are prorated for the time they served in each duty status. If a service member does not meet either the active or inactive requirement in a given month, flying hours attributed to one type of duty status can be applied to meet the requirement in the other. Additionally, there are several cases in which officers still qualify for AvIP but do not meet the flying requirements. We discuss these cases in the section on AvIP in this chapter.

[e] Applies to RC officers who are entitled to compensation under 37 U.S.C. § 206.

[f] Service members cannot receive HFP and IDP at the same time.

[g] 37 U.S.C. § 206 specifies that RC members receive compensation according to a proportion of basic pay that an AC member receives. This is somewhat misleading. *Compensation* typically refers to all monetary compensation that a service member receives, including basic pay, basic allowance for housing, and special pays (if applicable). In that regard, the monthly compensation, as specified by 37 U.S.C. § 206, that an RC member receives for IDT is more like basic pay because it is not also based on the additional compensation that an AC member receives. Sources are split as to whether they use an RC member's basic pay or compensation to calculate the amount of S&I pay (DoDI 1340.09, 2018, uses compensation; Office of the Under Secretary of Defense [Comptroller], 2020, uses basic pay). However, this is a formality: The monthly amount that an RC member receives does not change, regardless of whether the proportion of basic pay they receive every month is called compensation or basic pay.

[h] Service members cannot receive this HDIP if they are receiving other HDIPs concurrently.

[i] There is no language that precludes an RC member on IDT from receiving these HDIPs in either DoDI 1340.09, 2018, or Chapter 24 of Office of the Under Secretary of Defense (Comptroller), 2020. However, these HDIPs are not listed as a pay an RC member on IDT could receive in Chapter 58 of DoDI 1340.09, 2018; in practice, it is unlikely that an RC member on IDT would be eligible to receive these HDIPs.

[j] Office of the Under Secretary of Defense (Comptroller), 2020, Ch. 11, states that the maximum rate for diving duty is $340 for enlisted service members.

[k] Only service members who receive basic pay are eligible. Service members paid under 37 U.S.C. § 204 (generally, active duty) receive *basic pay*, whereas reservists generally receive *compensation* through 37 U.S.C. § 206.

Aviation Incentive Pay

AvIP is authorized for officers under 37 U.S.C. § 334 and for enlisted members who operate remotely piloted aircraft under 37 U.S.C. § 334a. For the purposes of this report, the term *AvIP* refers to both incentive pays. The policies and procedures for implementing AvIP are given in DoD Instruction (DoDI) 7730.67.

Officers are eligible if they have an aeronautical rating or are in training for an aeronautical rating and perform operational flying duty (OFD) or proficiency flying duty (PFD). They also must continuously complete a minimum number of flying hours. The minimum flight requirements for officers in the AC is four hours per month or 24 hours over six months. As shown in Table 2.1, the minimum flight requirement for RC officers is half the AC requirement (two hours per month or 12 hours over six months).

Some officers receive AvIP even if they do not meet the flying requirement or are not performing OFD or PFD. This occurs in the following circumstances. First, the service secretaries may waive these flying requirements for officers, except flight surgeons and medical officers, who are assigned to OFD and PFD positions and who otherwise meet the eligibility criteria. Second, service secretaries may also waive flying requirements in "extreme circumstances," e.g., when aircraft are unavailable or for military operations (Office of the Under Secretary of Defense [Comptroller], 2020). Finally, service secretaries may allow officers to receive AvIP if the officers have received AvIP for at least 12 years of aviation service (YAS) and serve in certain positions, such as a joint assignment or attending professional military education. The length of time that officers can qualify for continuous AvIP depends on the number of creditable months of OFD and PFD they previously completed. For example, if an officer completed 96 creditable months of OFD or PFD in their 12 YAS, they would be eligible to receive AvIP for up to 18 years.[2]

---

[2] There are several differences between the amount of AvIP and eligibility for AvIP among officers and warrant officers. Warrant officers can receive an amount of AvIP corresponding to ten YAS after they reach 22 YAS until they retire. Additionally, officers who no longer meet monthly flying requirements typically are no longer eligible to receive AvIP after 25 YAS. Warrant officers, on the other hand, typically can receive it until they retire as long as they meet the other eligibility criteria.

The maximum authorized amount of AvIP is $1,000, but this amount varies by YAS and service branch, from $125 per month to $1,000 per month (Defense Finance and Accounting Service, undated). Table 2.2 shows the AvIP pay rates by service and YAS. RC members are paid at a rate of one-thirtieth for each eligible training period of instruction during IDT. For example, an RC member in the Air Force with between six and ten YAS would qualify for $700 per month if serving in an active duty status. However, if they were in an inactive duty status and participated in exactly four weekend drills that month, they would receive four-thirtieths of $700, or $93 (DoDI 7730.67, 2016).

Hazardous Duty Pay

The policies and procedures for implementing hazardous duty pay are listed in DoDI 1340.09, which collectively terms the pays "hazard pay," or HzP.[3] The HzP program consolidated HFP, IDP, and multiple categories of HDIP to be paid under 37 U.S.C. § 351. We provide a brief overview of each of these pays in this section. As we describe later, policies with respect to proration are identical for AC and RC personnel for HFP and IDP. Thus, concern about unfairness of prorating HDP for RC members has focused on HDIP rather than HFP and IDP. We describe each of the categories of HDIP as well.

Hostile Fire Pay

Service members who perform duty in a hostile fire area or who are exposed to a hostile fire event are eligible for HFP at a rate of $225 per month. HFP is not prorated for either AC or RC members. In that respect, it is unlike the other S&I pays in this report. Exposure to or danger from a hostile fire event must be certified by the on-scene commander, and hostile fire areas must be designated by the service secretary. Once eligibility is established,

---

[3] The abbreviation *HDP* is not used for hazardous duty pay but rather for hardship duty pay, which compensates service members who are assigned to locations or performing missions in locations where the conditions are substantially below those expected in the continental United States.

TABLE 2.2
Aviation Incentive Pay Authorized Benefit Pay Rates

YAS	Maximum Authorized	Army	Navy	Navy Officers, Administrative Milestone Billets	Marine Corps	Air Force
Effective date	10/1/2017	1/1/2020	10/17/1998	4/1/2018	3/1/2018	10/1/2017
2 or fewer	$150	$125	$125	$125	$125	$150
More than 2	$250	$200	$156	$156	$156	$250
More than 3	$250	$200	$188	$188	$188	$250
More than 4	$250	$200	$206	$206	$206	$250
More than 5	$250	$200	$206	$206	$206	$250
More than 6	$800	$700	$650	$650	$650	$700
More than 7	$800	$700	$650	$650	$650	$700
More than 8	$800	$700	$650	$650	$800	$700
More than 9	$800	$700	$650	$650	$800	$700
More than 10	$1,000	$1,000	$650	$1,000	$1,000	$1,000
More than 11	$1,000	$1,000	$650	$1,000	$1,000	$1,000
More than 12	$1,000	$1,000	$650	$1,000	$1,000	$1,000
More than 13	$1,000	$1,000	$650	$1,000	$1,000	$1,000

Table 2.2—Continued

YAS	Maximum Authorized	Army	Navy	Navy Officers, Administrative Milestone Billets	Marine Corps	Air Force
Effective date	10/1/2017	1/1/2020	10/17/1998	4/1/2018	3/1/2018	10/1/2017
More than 14	$1,000	$1,000	$840	$1,000	$1,000	$1,000
More than 15	$1,000	$1,000	$840	$1,000	$1,000	$1,000
More than 16	$1,000	$1,000	$840	$1,000	$1,000	$1,000
More than 17	$1,000	$1,000	$840	$1,000	$840	$1,000
More than 18	$1,000	$1,000	$840	$1,000	$840	$1,000
More than 19	$1,000	$1,000	$840	$1,000	$840	$1,000
More than 20	$1,000	$1,000	$840	$1,000	$840	$1,000
More than 21	$1,000	$1,000	$840	$1,000	$840	$1,000
More than 22	$700	$700	$585	$700	$585	$700
More than 23	$700	$700	$495	$700	$495	$700
More than 24	$450	$400	$385	$450	$385	$450
More than 25	$450	$400	$250	$450	$250	$450

SOURCE: Defense Finance and Accounting Service, undated. Army rates prior to 2020 can be found in DoDI 1340.09, 2018.

the service member earns the full amount of $225.[4] RC members (like AC members) may be eligible for HFP even if they are exposed to a hostile fire event on U.S. soil. For example, RC members who were at the Pentagon on September 11, 2001, would have been eligible for HFP, no matter if they were on IDT, AT, or another duty status.

Imminent Danger Pay

Service members are eligible for IDP if they have been permanently assigned to or perform duty in an area that is designated eligible for IDP. The authorized amount is $225, which is prorated according to the number of days they were authorized to be in the area. Proration of IDP occurs for both AC and RC personnel. For example, if a service member performs duty in location B from June 1 to June 10, or ten days out of the month, they would receive one-third of IDP for that month (or $75). RC members on IDT are prohibited from performing drills in a location designated to receive IDP (DoDI 1215.06, 2015). Although RC members can perform AT in an imminent danger area, it is rare (DoDI 1215.06, 2015). For example, a pilot flying through IDP-designated airspace during AT would receive IDP for the day of the flight.

Hazardous Duty Incentive Pay

Service members performing one of 13 designated hazardous duties are eligible for HDIP. These duties range from parachute jumping to exposure to highly toxic pesticides. The authorized amount for HDIPs ranges from $150 per month to $250 per month depending on the hazardous duty. For most of the HDIP categories, AC members who maintain their qualifications for the full month receive the full amount unless their orders to perform the hazardous duty begin or end in the middle of a month. In these cases, their S&I pay is prorated at a rate of one-thirtieth for every day they received basic pay and were eligible. RC members performing annual training are eligible for one-thirtieth of the authorized amount for every day of AT (typically,

---

[4] DoDI 1340.09, 2018, describes the rules for establishing hostile fire events and hostile fire areas.

14 days). RC members performing IDT, on the other hand, receive a prorated amount based on their basic pay compensation for the month. For example, a flying RC crew member who performed four IDT drills in a month would receive compensation at a rate of four-thirtieths of the basic military pay for their pay grade and years of service and four-thirtieths of $250 (i.e., $33) for flying duty as a crew member that month (DoDI 1340.09, 2018).

Except for flying duty, all of the various HDIP categories have the same eligibility and training requirement for AC and RC personnel. For example, for HDIP for demolition duty, both AC and RC personnel must perform demolition using explosive objects, obstacles, or other explosives at least once per month. The difference between AC and RC personnel is in the proration of the pay. All members technically receive HDIP and AvIP at a prorated rate. For AC members, proration is based on days of qualifying service; for RC personnel, proration is based on either days (for active duty statuses) or drill periods (for inactive duty statuses). Although AC members may be more likely to earn the full monthly rate (because they perform qualifying service for a full month), RC members on IDT are actually receiving twice as much pay per qualifying day.[5]

The one exception is HDIP for flying duty. To be eligible for HDIP for flying duty, AC personnel must participate in at least four hours of aerial flight each month. In contrast, RC personnel must participate in two hours of aerial flight each month. As with other categories of HDIP, the authorized benefit for RC personnel is prorated, unlike the benefit for AC personnel.

The current guidance for implementing HDIP dates to January 2018. Prior to that, HDIP was paid under 37 U.S.C. § 301 and 304 rather than 37 U.S.C. § 351. As noted in Asch, Marrone, and Mattock, 2019, the change had several implications for HDIP rates, with pay structure being most relevant to this report. Previously, pay rates sometimes varied by pay grade and years of experience. Now, pay rates are standardized.

Other changes accompanied the consolidation of hazardous duty pays under section 351 that may have affected whether a service branch actually

---

[5] We note that AC members who are available for only a partial month because they rotate in or out of an assignment during the month will also receive prorated HDIP for flying duty based on the number of days in the month in the assignment. This is true of all HzP, with the exception of HFP.

pays HDIP in practice. For example, at the same time as HzPs were consolidated, a master diver critical skill incentive pay was authorized under 37 U.S.C. § 353. This pay has similar eligibility requirements as HDIP for diving duty, but a service member cannot earn both concurrently. Thus, if a service chooses to implement the pay under section 353, then it would not apply to this report.

A further change to the HzP program was instituted in the FY 2021 NDAA (Pub. L. 116-283). Section 613 raised the maximum allowable monthly rate from $250 to $275 for IDP and all HDIPs. However, the statutory maximum is not the same as the full monthly rate in practice; currently, only HDIP for flying duty for crew members is paid at $250 for a full month. The NDAA also amends the statutory language to refine how HzP may be prorated. Instead of mandating that the services prorate IDP and HDIP, section 614 allows service secretaries to prorate and, notably, allows them to remove the proration requirement for individual pays. Although none of these statutory changes have thus far resulted in a change in policy, they could affect the way these pays are implemented in the future.

Pays Included in Analysis

Table 2.1 and the discussion in this chapter show that some of the special pays identified by the NDAA are implemented the same for RC and AC members: For HFP and IDP, there are no differences in eligibility and no differences in proration.[6] Further, some of these pays cannot be earned by RC members in a training duty status, although they may be eligible in other active duty statuses. Finally, after investigating the data (described in Chapter Three), we found that some pays are not actually earned by RC members, though existing authority indicates that they could be eligible. Therefore, our analysis is limited to those pays that RC members actually earn and for which RC and AC eligibility and/or payment actually differs.

---

[6]   There are no differences in practice, since RC members are not eligible for IDP when in an inactive status. Therefore, the distinction between prorating per day versus per drill period is not germane for that pay.

Table 2.3 summarizes the pays that are included and excluded from our analysis. HFP and IDP are excluded because there are no differences in those pays for RC and AC members. HDIP for WMDCS is excluded because RC members cannot earn that pay in a training status: They must be on active duty for at least 139 days to attain eligibility. Five HDIPs are excluded because we did not actually observe RC members earning those pays during our observation window: polar region operations, chemical munitions,

TABLE 2.3

Special Pays Included and Excluded from Report

Pay	Notes
Included in Cost Calculations	
• AvIP • HDIP, flying duty (crew) • HDIP, flying duty (noncrew)	• RC and AC eligibility differ in flying hours.
• HDIP, parachute duty (static) • HDIP, parachute duty (HALO) • HDIP, demolition duty • HDIP, experimental stress duty • HDIP, flight deck hazardous duty • HDIP, duty involving toxic fuels and propellants	• RC and AC eligibility are the same.
• HDIP, diving duty	• RC members are ineligible during IDT.
Excluded from Cost Calculations	
• HFP • IDP	• No differences in eligibility or pay amounts between AC and RC.
• HDIP, WMDCS team	• RC members are ineligible when in a training duty status.
Not Observed in Data	
• HDIP, polar region flight operations duty • HDIP, duty involving handling chemical munitions • HDIP, maritime VBSS duty • HDIP, duty involving exposure to highly toxic pesticides • HDIP, laboratory duty utilizing live dangerous viruses or bacteria	• These items are excluded from calculations because no RC members were observed earning these pays.

SOURCES: DoDI 1340.09, 2018; Office of the Under Secretary of Defense (Comptroller), 2020, Ch. 11, 22, and 24; and data provided by DMDC.

NOTE: HALO = high altitude, low opening.

maritime VBSS, toxic pesticides, and virus lab duty. The other pays are included in the analysis. Note that RC members are not eligible for HDIP for diving duty while on an inactive duty status, but they can earn the pay while on an active duty status.

Reserve Component Structure

The archetypal RC member might be a reservist who serves on IDT one weekend per month and on AT for two weeks each year, known as a *drilling reservist*. But drilling reservists comprise just one subset of the RC. Other subsets of the RC (for example, aviation personnel) may see considerably greater rates of participation, with some RC aviators having 72 additional drill periods, for a total of 120 drill periods plus annual training plus any other active duty they may perform.[7] Changing the payment method for special pays would affect anyone who earns those pays, even if they do not earn them regularly. Pay amounts also depend on the duty status being performed; RC members could, in theory, earn these pays by performing duty statuses other than IDT or AT. Therefore, it is important to identify who *can* earn special pays and in what types of duty status.

Duty Status

To earn special pays, a service member must earn basic pay under either section 204 or section 206, as described in Appendix A. Special pays are pro-rated according to the amount of basic pay, meaning that training without pay, or compensation paid under a different authority, will not be counted toward special pay. Duty statuses are performed according to various utilization categories, which are described in DoDI 1215.06. These utilization categories are training, support, mobilization, and other. Figure 2.1 summarizes the statuses corresponding to training and support. Under a mobilization status, the RC member typically would be supplementing AC forces, so mobilizations are excluded from this study. Funeral honors duty

---

[7]   We thank Tom Bush, former principal director for manpower and personnel within the Office of the Assistant Secretary of Defense for Reserve Affairs, for this example.

FIGURE 2.1

Reserve Component Duty Statuses for Training and Support

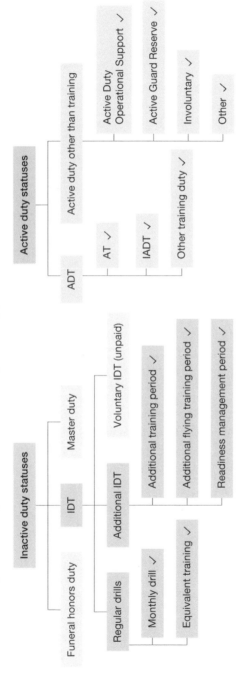

SOURCE: DoDI 1215.06, 2015.

NOTE: ADT = active duty for training; IADT = initial active duty for training. Check marks indicate that RC members may earn special pays when serving in that duty status. National Guard components use slightly different acronyms. For a comparison, see DoDI 1215.06, 2015, Enclosures 3 and 4.

and muster duty fall in the "other" category. The DMDC data include payments for both of these, but we ignore them because we cannot always verify the source of compensation and because we did not observe the number of periods paid.[8] All other statuses are ignored in this study because they are not recorded in DMDC's master file for RC personnel, known as the Reserve Active Duty Transaction File; they correspond to medical care, disciplinary actions, captive status, or other types of involuntary service.

A key conclusion from Figure 2.1 is that any assessment of special pays must account for more than just IDT and AT. RC members can perform hazardous duties on other training statuses (particularly, additional flight training periods, or AFTPs, which are specifically for flight crew training), and periods served in a support status will still count toward the proration baseline even though hazardous duties are not likely to be performed under such a duty status. For example, suppose an RC member does 14 days of AT, during which they fulfill a parachute jump requirement, then performs an additional three days of Active Duty Operational Support duty, during which they do not jump. For that month, HDIP for parachute duty would be prorated at seventeen-thirtieths of a full month, not fourteen-thirtieths, as implied by AT alone.

Reserve Component and Training and Retirement Categories

An RC member's RC category (RCC) and training and retirement category (TRC) determine whether they can train for pay and therefore earn special pays. Every RC member not counted toward AC end strength is placed to serve in one of several RCC/TRC combinations, which are described in

---

[8] Muster duty is compensated under per diem rules under 37 U.S.C. § 433 and therefore would not contribute to the "compensation" against which special pays are prorated. Funeral honors duty might be unpaid, or compensation might be paid as an allowance under 37 U.S.C. § 495 or as compensation under section 206. When paid under section 206, funeral honors duty could, in theory, count toward the proration of special pays. However, in practice, ignoring it will not affect our estimates: We observed 1,443 instances of funeral honors duty being paid, and only two instances in which it was paid in the same month as one of the special pays studied in this report.

DoDI 1215.06. Only a subset of RCC/TRC codes are eligible for training with pay.

Figure 2.2 shows the RCC/TRC codes for the Ready Reserve, which includes the Selected Reserve, Individual Ready Reserve, and Inactive Guard. (The RC also includes the Standby Reserve, which does not train for pay, and the Retired Reserve.) Drilling reservists are those who are most likely to perform the stereotypical weekend of monthly IDT plus two weeks of annual AT, and these members are highlighted in orange with RCC/TRC code "SA."

Members in several other categories may train for pay. The categories highlighted in purple are included in our analysis: military technicians not assigned to units, Individual Mobilization Augmentees, RC members in the training pipeline, trained members of the Individual Ready Reserve, and a limited set of members of the Inactive Guard. These members may earn special pays if and when they train, although they might train infrequently, if at all.

Categories in green are excluded from this study, either because they are not allowed to train for pay or because, like those on Active Guard Reserve duty, they are considered full-time support.

A key conclusion from Figure 2.2 is that any analysis of special pays must include more than just drilling reservists. Although most special pay costs might be associated with drilling reservists, a full estimate of special pay costs needs to include all eligible RC members.

FIGURE 2.2

Reserve Component Categories and Training and Retirement Codes for the Ready Reserve

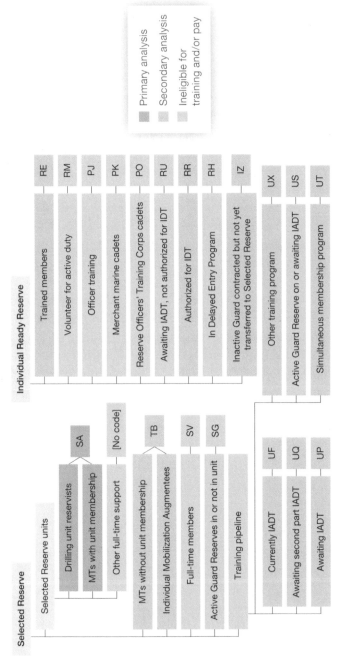

SOURCE: DoDI 1215.06, 2015.

NOTE: MT = military technician. The two-letter codes indicate the RCC code (first letter) and TRC code (second letter).

CHAPTER THREE

Data and Research Approach

To conduct our analysis to respond to the two issues that Congress raised regarding cost and potential qualification for HDIP, we require three types of data. First, we need data on the amount of service that RC members provided: specifically, data on the amount and type of duty each RC member performed in an inactive duty status for training that resulted in payment of the S&I pays of interest. This information will tell us how much duty was performed that resulted in S&I pay.

Second, we require data on the amount that RC members were paid for the duty that was performed (by type of duty) based on the current proration policy. As we describe in this chapter, this information will provide a baseline of payment under current policy when we conduct simulations of how much these members would be paid instead under a full-rate policy. Furthermore, it is important for us to align the dates on when the duty was performed with payments so we know which payments correspond with which duty.

Third, because AvIP and HDIP for flying depend on the number of flying hours, and the rules on the number of flying hours required to be eligible for these pays differ for RC versus AC members (as described in the previous chapter), we also require data on flying hours.[1]

As we describe in this chapter, available data do not meet these requirements. Consequently, we developed a methodology—or, more accurately, three alternative methods—to make the best use of the data at hand.

We begin the chapter by discussing our data sources and the construction of our analysis file from these sources. Next, we discuss the quality of

---

[1] AvIP may be paid to officers who do not perform flying duty if they qualify for "continuous AvIP."

the data from the standpoint of our data requirements. Finally, we describe our methodology for estimating the cost under a full-rate policy, given the drawbacks of the data. We present our results in Chapters Four and Five.

Data Sources

DMDC Data on Pay and Duty Status

From DMDC, we accessed pay records and duty status for all RC members who served from January 2018 to March 2020. These dates were chosen so that we could track how special pays were disbursed from the beginning of the current HzP program in 2018 until the beginning of closures in response to the coronavirus disease 2019 (COVID-19) pandemic in March 2020. After March 2020, DoD temporarily waived the performance requirement for the payment of performance-based pays because of the pandemic and service members' inability to perform those duties through no fault of their own (DoD, 2020).

One of the key challenges we faced with respect to accessing the required data was that pay and duty status information are recorded in different sources. (We discuss why this is challenging later in the chapter.) Pay information is recorded monthly and includes the amount of basic pay for active and inactive duty received that month, as well as the amount of each special pay received (see DoDI 7770.02, 2019). The pay file disaggregates basic pay for inactive duty by listing the number of drill periods paid for each type of IDT; it does not disaggregate basic pay for active duty in the same way. The pay file also provides administrative information, such as whether the pay comes from AC or RC sources and whether a member's pay account is active.[2] In short, the pay file tells us how much basic pay and S&I pay an RC member received, the pay's AC versus RC source, and the number of drill periods paid. It does not tell us the amount of active duty periods paid and the type of active duty performed.

---

[2]   In theory, all RC members should be paid from RC pay accounts. However, sometimes the data are ambiguous regarding a service member's component; in other cases, records may be outdated. See Appendix C for more information on how we cleaned the data.

Instead, active duty status information comes from a different data source. Specifically, it comes from the RC Active Service Transaction File (see DoD Manual 7730.54, Volume 1, 2019), which lists each stint of active duty service for each RC member. Stints are listed with start and end dates, associated statutory authorities, and project codes. The statutory authorities and project codes distinguish different types of ADT statuses, active duty other than training statuses, contingency operations, and disaster relief operations. From this information, we calculated how many days in each month each RC member served in each possible active duty status. However, the transaction file does not list dates of inactive duty statuses; the pay file is the only source of information about inactive duty service for RC members.

We supplemented pay and service data with additional information from DMDC personnel files indicating their RCC/TRC code, occupational assignment code, and information indicating deployment status.

Reserve Component Information on Flying Hours

Individual RCs provided aggregated data on RC members' flight hours. The DMDC data do not track individual flying hours, which is one criterion by which RC and AC eligibility differs for AvIP and HDIP for flying duty. If eligibility requirements for RC and AC members were made the same, then RC members who fly at least two but less than four hours in a month would lose eligibility for these pays.

Therefore, on behalf of our study, the Office of the Secretary of Defense requested that each RC provide annual aggregate counts of the number of members who earned AvIP or HDIP for flying duty based on how much these members flew. We requested separate counts for each pay, for each of the following three categories:

1. RC members who earned the pay but flew less than 24 hours in a year; these members are most likely earning continuous AvIP or earned HDIP for only a fraction of a year.
2. RC members who earned the pay and flew at least 24 but less than 48 hours in a year; these members are most likely to lose eligibility under a policy change that would realign the flight hour require-

ment to be in accordance with the AC (four hours rather than two hours).

3. RC members who earned the pay and flew more than 48 hours in a year; these members would maintain eligibility under a policy change.

For each of the three groups, we also requested a count of the total periods of IDT and days of AT served by RC members in that group each year.

Analytic Sample Construction

Using the guidance in DoDI 1215.06 and the service-related information from DMDC, we selected a subsample from the DMDC data for analysis. The subsample included RC members in each month who (1) had a pay account in good standing that was being paid out of RC funds and (2) were theoretically eligible to earn the special pays that we studied.

To identify RC members receiving RC pay, we examined the sources of a member's pay and benefits, as well as their duty status. We excluded members when the pay data indicated that their pay account was linked to AC funds or when the member was listed during the same month in the AC personnel file. We also excluded members whose accounts were suspended or closed. We then excluded months during which an RC member was deployed and months during which they were eligible for full-time active duty health benefits. These criteria limited the sample to months during which RC members were available to serve in a training or support status, as illustrated in Figure 2.1.

To identify RC members who were theoretically eligible to earn special pays during training, we excluded RCC/TRC codes that were not eligible to drill with pay, as described in Figure 2.2. Members in the Individual Ready Reserve were included only if their personnel file indicated they were authorized to perform IDT. Appendix C provides more information on how we constructed the sample.

Table 3.1 summarizes the resulting analytic data set. The sample includes 935,080 RC members, observed over a total of 18,571,700 months. Drilling reservists account for at least 70 percent of the sample in every RC and

TABLE 3.1

Reserve Component Members Eligible for Special Pays During Training, January 2018 to March 2020

RC	Individual RC Members			Total Observations (person-months)		
	Drilling Reservists	Other RCC/ TRC Code	Total	Drilling Reservists	Other RCC/ TRC Code	Total
Army National Guard	281,190	113,981	395,171	6,699,873	1,106,630	7,806,503
	(71%)	(29%)		(86%)	(14%)	
Army Reserve	176,512	47,617	224,129	4,084,859	453,038	4,537,897
	(79%)	(21%)		(90%)	(10%)	
Coast Guard Reserve	7,135	2,841	9,976	143,920	37,329	181,249
	(72%)	(28%)		(79%)	(21%)	
Air National Guard	92,217	17,675	109,892	2,140,299	169,698	2,309,997
	(84%)	(16%)		(93%)	(7%)	
Air Force Reserve	62,306	19,499	81,805	1,424,424	296,296	1,720,720
	(76%)	(24%)		(83%)	(17%)	
Marine Corps Reserve	44,973	3,431	48,404	714,698	62,001	776,699
	(93%)	(7%)		(92%)	(8%)	
Navy Reserve	61,041	4,662	65,703	1,137,571	101,064	1,238,635
	(93%)	(7%)		(92%)	(8%)	
Total	725,374	209,706	935,080	16,345,644	2,226,056	18,571,700
	(78%)	(22%)		(88%)	(12%)	

SOURCE: RC Active Service Transaction File from DMDC. See Appendix C for more information on sample construction.

NOTE: The numbers in this table do not indicate the end strength for each component. They differ for two potential reasons. First, the analysis is a look over three years, so there is some churn or turnover. Second, we are including people who do not count toward end strength (such as military technicians), but their pay comes from the RC account, so S&I pay policy affects them.

78 percent of the sample in aggregate. The Army components are largest (a combined 66 percent of all RC members in the sample), followed by the two Air Force components (22 percent combined).

After identifying the sample of eligible RC members, we used the pay file and RC Active Service Transaction File to construct a panel data set, recording each RC member's service and pay each month. We constructed variables to record the total periods of active duty status in each month (by type of status), the total periods of IDT paid in each month, whether an RC member earned each special pay in each month, and the amount of each special pay.

Data Challenges

The available data have three key limitations for our purposes, as follows:

1. They do not provide complete information on periods of service performed by RC members; in some cases, we must infer the timing of service performed based on when RC members are paid.
2. Even when information on service performed is available, the data source on service performed differs from the source of information on payments for that service.
3. The payments of S&I pay can be "lumpy," meaning multiple months of service may be paid at once.

The result of these limitations is that we do not know the exact number of months an RC member performed the duty that qualified for hazardous duty pay, as we describe next. Therefore, we cannot estimate precisely how much an RC member should be paid if they received the full rate of special pay for each month that they performed the hazardous duty. We describe later in the chapter how we infer this information.

Comparing DMDC Data on Service Versus Payment

RC members must perform hazardous duties to earn special pays (except for continuous AvIP), but the actual duties are not observable in the data. We observe evidence that those duties were performed only in the form of a special pay disbursement. The challenge for us is that it is possible that an RC member serves in one month (i.e., pay is earned in one month) but is not paid until a later month. Table 3.2 shows an example in which an RC member might have served on AT over the course of two months but received the pay over the course of three months. In this example, the RC member performed their regular 14 days of AT at the end of July and through the first day of August. They qualified for HDIP for flying duty as a crew member in each month. The amount of HDIP they received is prorated based on the total periods of basic pay they received. In August, for example, they were paid six periods of basic pay (one day of active duty and five drill periods of IDT) and received six periods' worth of HDIP. However, the timing of basic pay does not align with the timing of service. In particular, the payment for AT is spread over three months, even though AT was served across two months. Therefore, the pay file implies a different pattern of service than the RC Active Service Transaction File implies.

Therefore, dates of service can tell us *in theory* when a pay might be earned, and pay files tell us *in practice* when it is paid out. Both are impor-

TABLE 3.2

Example of Differences in Dates of Service and Dates of Pay

Month	Information from RC Active Service Transaction File		Information from Reserve Pay File			Notes
	AT Days Served	Other Active Duty Days Served	Active Duty Days Paid	IDT Drills Paid	HDIP for Flying Duty (Crew) Periods Paid	
July 2018	13	0	1	0	1	AT overlaps the end of July and beginning of August. Only one day appears to be paid in July.
August 2018	1	0	1	5	6	The single day of August AT is paid in August.
September 2018	0	0	12	4	16	The remaining July days of AT are paid in September.

NOTE: Table shows a notional example based on actual patterns observed in DMDC data.

tant for our analysis, but they are not always observable for every relevant duty status. Table 3.3 illustrates whether periods can be observed or calculated from available information for both inactive and active duty statuses.

The table highlights two critical points. First, active duty payment and service information come from different files. We can estimate the number of months of eligibility for special pays according to when an RC member performed active duty. We can also calculate the number of days for which they were paid for active duty each month. But the two are not necessarily the same: Payment might not be disbursed in the same month in which service was performed, and payment might be lumpy, in that multiple months of service may be paid at once. Without the ability to align a payment with a particular date of service, we cannot know the exact number of months in which an RC member actually performed the hazardous duty. This means that we cannot know exactly how much an RC member should be paid if they received the full rate of special pay for each month they performed the hazardous duty.

Second, information regarding inactive duty comes only from pay data. The precise dates of IDT service are unknown and cannot be calculated from other variables. It is possible that payment for IDT, like payment for active duty, does not correspond perfectly to the periods served in a month. However, we cannot know the extent to which that is true. Therefore, our analysis conflates IDT drills paid and IDT drills served in each month.

TABLE 3.3

Information Sources for Service and Payment, by Duty Type

	Active Duty	Inactive Duty
Periods paid	• **Not observed:** Must be calculated by dividing total basic pay for active duty by daily basic pay rate. • Source: Reserve Pay File.	• **Observed:** Drill periods paid are listed for each type of IDT status. • Source: Reserve Pay File.
Periods served	• **Observed:** Days served are listed for each type of active duty status. • Source: RC Active Service Transaction File.	• **Not observed:** Cannot be observed or calculated from available data. • Source: None.

These data limitations mean it is not possible to calculate exactly how many months of full pay an RC member should earn under the full-rate policy. They do not, however, mean the data are wrong or that the pays are being implemented incorrectly. For example, the "lumpiness" of HDIP for flying duty might occur because multiple months' worth of pay can be earned in just one month. An RC member who does not fly enough hours in January and February could make up those hours in March if the member enters a three-month grace period wherein future hours flown can be retroactively applied to earlier monthly requirements.[3] An RC member can similarly qualify for multiple months of HDIP for parachute duty by performing parachute jumps in just one month.

The following methodology section provides an example and describes how we make use of the available data to produce a range of cost estimates. Appendix E provides more information on data challenges we faced.

Reserve Component Data on Flying Hours

All RCs responded to our requests for flying hour data. In some cases, our request posed a burden because of the way that personnel data systems recorded flight data. Table 3.4 shows the data that were provided for the project. We received information about AvIP from all services. We could not obtain information regarding HDIP, either because of system limitations in certain components or because the information was not sufficiently comprehensive to incorporate into the analysis.

We used the flying hours data available to estimate how many RC members would lose eligibility for AvIP or HDIP for flying duty if the monthly flight hours requirement were to change.[4] We do not have enough information to estimate how such a change would affect cost, since information on IDT drills and AT days was generally unavailable.

---

[3] Office of the Under Secretary of Defense (Comptroller), 2020, Ch. 22, provides several other examples of how AvIPs can be earned by flying various combinations of hours over multiple months.

[4] There is no indication that the flying hours requirement would change in the FY 2020 NDAA; the reporting requirement simply states "so long as there is at least one such period of instruction or appropriate duty in the month." It does not state that the AC and RC flight hour requirements would be aligned.

TABLE 3.4

Reserve Component Flying Hours Data

Component	AvIP		Notes
	Flying Hours	Drills and AT Days	
Army (Reserve and Guard)	✓	✓	AvIP data processed as requested. System limitations prevented delivery of HDIP information.
Air Force (Reserve and Guard)	✓		Received permission to use Air Force data housed at the RAND Corporation to calculate flying hours; duty status information not available.
Navy	✓	✓	AvIP data processed as requested. System limitations prevented delivery of HDIP information.
Marine Corps	✓	✓	AvIP data processed as requested. HDIP data not comprehensive enough for analysis.

NOTE: Check marks indicate the information was provided to us as requested. We did not request data from the U.S. Coast Guard Reserve.

Methodology

Calculating Annual Costs of Paying the Full Rate of Special Pays

The FY 2020 NDAA calls for an estimate of the yearly costs of paying RC members the full rate of a special pay for every month in which they perform hazardous duties. This estimation requires two steps. The first is to determine what a full months' pay should be. The second is to estimate how many months an RC member was eligible for each pay. Because of the data limitations discussed in the previous section, we developed three alternative ways of doing the second step.

Determining the Full Rate of Each Special Pay

We calculated the full rate for each pay using the eligibility requirements listed in DoDI 1340.09 and the information in the DoD Financial Management Regulation (see Tables 2.1 and 2.2). For most pays, the monthly rate does not vary by pay grade or years of experience. The exceptions during the period we studied are AvIP, HDIP for flying duty for crew members, and HDIP for diving duty.[5]

Using the tables of pay rates, we calculated the full rate of each special pay for each RC member in each month. Throughout this report, we refer to these rates as the *counterfactuals*: that is, how much someone would earn under the alternative policy if they were to fulfill the eligibility requirements for a pay in a given month. We also calculated the per-period rate of each pay by dividing the full monthly rate by 30. This is the amount that the RC member would earn for each day of active duty or drill period of IDT under the current special pay proration policy.

Determining Months of Eligible Service

After calculating the counterfactual full rates, we needed to develop a methodology for determining how many months an RC member was eligible for each pay. Given the data quality limitations described earlier, there is no precise way to do this. We can use dates of active duty service but remain unsure about whether the RC member actually performed the hazardous duty on those dates. Alternatively, we can use observed payments but remain unsure about whether those payments reflect more than one month of service.

We developed three cost estimation methods that variously leverage the data on service provided, the pay data, and a mixture of the two. The methods count the number of eligible months as follows:

- **Method 1 (data on service provided).** Use observed dates of service to count the number of months of eligibility each year. Conditional on earning the pay during a given calendar year, count each month in which an RC member performs active duty service or in which they are paid for inactive duty service as one month of eligibility. The advantage of this method is that the computation is directly related to what we are attempting to measure (the amount of duty performed). The

---

[5] Specific rates for HDIP for diving duty are found in Office of the Under Secretary of Defense (Comptroller), 2020, Ch. 11, "Special Pay – Diving Duty." Rates for flying duty varied by pay grade from 1998 through January 2020, at which time the Financial Management Regulation was updated to standardize the full rate to $250. See Office of the Under Secretary of Defense (Comptroller), 2020, Ch. 22, "Aviation Incentive Pays."

disadvantages are that we do not have complete data on the service provided and the dates of service provided do not perfectly align with the dates of payment.

- **Method 2 (pay data).** Use observed payments to count the number of months of eligibility each year. Count each month in which a special pay is disbursed as one month of eligibility. The advantage of this approach is that we can directly observe disbursements of S&I pay. The disadvantages are that we are unable to align any payment with the amount of duty performed and pays may be lumpy.
- **Method 3 (benchmarking periods paid against periods served).** For each month, compare the estimate of the number of periods served (used in Method 1) with an estimate of the number of periods paid (used in Method 2). If periods served equals periods paid, assume the pay represents one month of eligibility. If they are not equal, assume the pay accounts retroactively for past service and compute months as equal to periods paid divided by average periods served. To make this computation, we must compute the average number of periods served and the total number of periods paid. We compute the average periods served as the total periods served (equal to active duty days served plus inactive drill periods paid) divided by months when the RC member performed any service. The number of periods paid is estimated as the total amount of special pay disbursed divided by the per-period rate. The advantage of this method is that it accounts for the lumpiness of pay over time because it benchmarks periods paid to periods served and inflates or deflates each pay disbursement according to how it compares with observed patterns of service. The disadvantage is that the benchmarking approach is ad hoc.

After determining the number of months of eligibility for each RC member each year, we multiply by the full monthly rate of that pay for that RC member to get an annual cost for each person. We then sum over all individual members of each RC to attain an aggregate annual cost estimate.

Each method might yield a different estimate of the total number of eligible months, and each estimate could differ from the true number of months (which is unobserved). Table 3.5 provides a notional example. The example shows an RC member who sometimes earns HDIP for demolition duty.

TABLE 3.5
Notional Example of Three Methods for Calculating Special Pay Eligibility

Month	Demolition Duty Performed? (Unobserved)	Observed Service			Observed Payment	Calculated Eligibility		
		Active Duty Days Served	IDT Drills Paid	Total Periods "Served"	HDIP for Demolition Duty Periods Paid	Method 1: Count Months with Service	Method 2: Count Months with HDIP Payment	Method 3: Compare Periods Paid with Periods Served
January 2019	Yes	7	0	7	0	1	0	0
February 2019	Yes	7	0	7	14	1	1	14 / 6 = 2.3
March 2019	Yes	0	5	5	5	1	1	1
April 2019	Yes	1	4	5	0	1	0	0
May 2019	No	0	6	6	5	1	1	5 / 6 = 0.8
Notes	True months of eligibility: 4	Average periods served: 6				Method 1 result: 5 eligible months	Method 2 result: 3 eligible months	Method 3 result: 4.1 eligible months (round up to 5)

This RC member serves either active duty or IDT every month for five months, averaging six periods of service each month. They actually perform demolition duty (blue column) only four out of five months, but this is not observable in our data. Instead, we observe the amount of service (purple columns). The member serves 14 days of AT spread over January and February, followed by typical IDT drill weekends with four to six drills in each of the next three months. In April, they also perform three days of active duty in a support capacity (such as Active Duty Operational Support).

We also observe how much HDIP for demolition duty the RC member earns each month (green column). The payments do not align with service. The demolition duty performed during AT is compensated by one aggregate payment in February. Demolition duty from the March IDT weekend is compensated in March, but demolition duty for the April IDT weekend is not paid out until May. Because we do not actually observe the hazardous duty, we cannot determine whether the May payment reflects a duty performed in April or a partial payment for the duty performed in May.

Each of the three eligibility calculations arrives at a different result. The RC member should actually earn four months of full HDIP for their performance of hazardous duties. Method 1, using the service data alone, counts five months of eligibility. Method 2, using pay data alone, counts three months of eligibility. Method 3, which compares payment and service, arrives at just more than four months of eligibility, which is rounded to five.

In this example, the method based on observed payment led to the smallest count of eligible months, and the method based on observed service led to the largest estimate. In general, this need not be true. Different methods might yield the highest or lowest estimate depending on the way payments and service are observed over time. For example, if pay is particularly lumpy, such that payments are made infrequently for several months at a time, counting eligibility using months of payment will yield a relatively small count. Conversely, if payments are made piecemeal, such that a month's worth of service is spread across several paychecks, then counting eligibility using months of payment will yield a relatively high count.

Because our estimates of special pay eligibility are necessarily imprecise, the cost estimates presented in Chapter Four should be considered as approximations of the lower bound on the actual costs of an alternative special pay policy. In some cases, our cost estimates suggest a *lower* total amount than is currently paid out. This occurs when payments are made infrequently in large amounts. As an extreme example, if an RC member were paid just once in a year with a large amount that covered the past 12 months of service, then Method 2 would count that payment as one month of eligibility. The total implied cost of one full month could be less than the actual amount of that single pay disbursement. We do not report estimates in those cases in which the full-rate cost estimates are lower than the current annual costs.

Counting the Eligible Reserve Component Population

The FY 2020 NDAA also calls for an estimate of the number of RC members who qualify or potentially qualify for special pays according to their professions or required duties. To obtain those estimates, we used two approaches.

First, we calculated the average number of RC members who receive each special pay in a month. This provides an estimate of the number of RC members who qualify for a pay in a typical month.

Second, we calculated potential eligibility using the duty occupations that are associated with each HDIP. We grouped each service's primary duty occupation codes into categories, using the first two characters of the Army military occupational specialty code or Air Force Personnel Center code, or the full Navy and Marine Corps occupational codes. We then tabulated the primary duty occupational categories for RC members who earned each HDIP. Using those tabulations, we identified duty occupation categories that accounted for a large proportion of those who earned the pay. We treated these occupation categories as *eligible* categories. Finally, we calculated the average number of RC members assigned to those eligible categories in a month. That number is our estimate of the typical eligible population.

Table 3.6 provides an example of our occupational eligibility calculation using HDIP for flying duty for crew members in the Army Reserve. These calculations are from our analytic data set. We find that 12 occupational categories earned HDIP for flying duty for crew members at least once during our observation window. But just two of those occupational categories account for more than 99 percent of all payments in an average month: aviation occupations and those in the medical career management field. We treat these two categories as the eligible occupations for Army Reserve members to earn this pay.

TABLE 3.6

Example of Occupational Eligibility Calculations (HDIP for Flying Duty for Crew Members Among Army Reserve)

Army Duty Occupation Category	Average Number of Payees per Month	Percentage of Monthly Payees	Average Number of RC Members Assigned to Duty Occupation Category Each Month
Undesignated Occupation	<1	<1	13,101
Technical Specialists	<1	<1	3,547
Interpreter/Translator	<1	<1	3,072
Infantry	<1	<1	1,825
Engineer	<1	<1	14,782
Aviation	**82.5**	**87.8**	**1,953**
Psychological Operations	<1	<1	2,299
Medical	**10.7**	**11.4**	**13,344**
Mechanics/Maintenance	<1	<1	13,588
Quartermaster Corps	<1	<1	22,000
Unknown	<1	<1	1,531
Total	**94**	**100**	**91,042**

SOURCE: DMDC data.

NOTE: Rows in bold show occupation categories that are considered eligible for HDIP for flying duty for crew members.

The sum of the first column is equal to our first estimate of eligibility: the average number of payments disbursed each month (in this case, 94). Adding the bold numbers in the last column provides our second estimate of eligibility: There are 15,297 Army Reserve members who are *potentially* eligible each month. The potentially eligible number is far greater than the actual number of payments, a pattern we find in general among all pays and that we discuss in more detail in Chapter Five.

For AvIP, we did not have to calculate eligible occupations. AvIP is paid according to YAS; therefore, to calculate potential eligibility, we counted the average number of RC members with nonzero YAS.

CHAPTER FOUR

Cost of Paying Reserve Component Members the Full Rate of Special and Incentive Pays

In this chapter, we describe our cost estimates for paying the full rate of special pays. We first describe who earns these pays and how much the pays cost under the current proration policy. We then describe the baseline results using RC flying hours data. Then we provide the cost estimates under a full-rate policy, calculated using each of the three methods explained in Chapter Three. Finally, we provide some evidence for why the increase in costs (compared with the current policy) is lower than we predicted for a stereotypical RC training schedule of four drill periods of IDT each month and two weeks of annual training.

Special and Incentive Pays Under the Current Proration Policy

Using the analytic data set described in Chapter Three, we calculated the number of RC members in each RC who earn each special pay in an average month. Figure 4.1 shows the average monthly earners for each HDIP. The most common pays are flying duty (for both crew and noncrew members), static line parachute duty, and demolition duty.

The two Army components have disproportionately large shares of HDIP earners compared with their overall size. The Army accounts for 67 percent of RC members in our sample but accounts for nearly 100 percent of RC members who earn parachute duty and about 75 percent of crew members

FIGURE 4.1

Average Monthly Number of Reserve Component Members with HDIP Payment, by Pay and Component

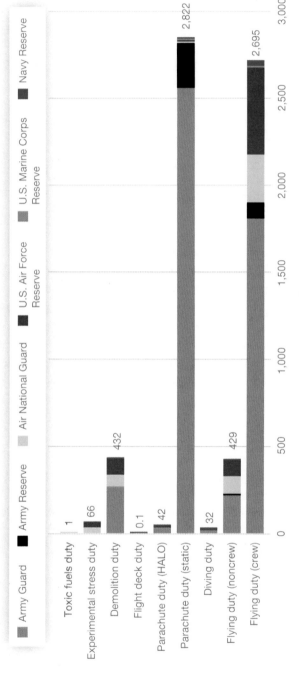

SOURCE: DMDC data, January 2018 to March 2020.

NOTE: The U.S. Coast Guard Reserve had no reported payments of any HDIP during the observation window.

who earn flying duty. The Air Force components account for most of the remaining payees, and the other components account for relatively few. The U.S. Coast Guard Reserve did not report any HDIP payments during our observation window.

Figure 4.2 shows the average number of RC members who are paid AvIP each month. Here, the Air Force components have disproportionately large shares of payees. The Air Force Reserve and Air National Guard each account for just over 25 percent of AvIP payees, whereas they have roughly 10 percent of RC members overall. The Navy Reserve and Army National Guard also have disproportionately large numbers of AvIP payees.

To provide information on the baseline cost of HDIP for RC members under the current policy of prorating S&I pay, we calculated the average annual costs to DoD that are associated with each pay under the current system. The results are shown in Figures 4.3 and 4.4 for HDIP and AvIP, respectively. Figure 4.3 shows that the proportions of costs of HDIP that are attributable to each RC are similar to those shown in Figure 4.1. The combined cost of all HDIPs is $6.8 million per year. Static parachute duty and crew member flying duty account for the majority, at more than $2.5 million each. Flying duty for noncrew members, experimental stress duty, and demolition duty each cost a few hundred thousand dollars, and the others each cost less than $100,000 per year.

Figure 4.4. shows the annual costs of AvIP, which far outpace the aggregate costs of HDIPs. The aggregate costs of AvIP are $38.8 million per year, roughly six times the cost of all HDIPs combined. The proportional costs for each RC are similar to the proportions of earners shown in Figure 4.2. The Army National Guard and both Air Force components each pay more than $10 million in AvIP per year, more than three times the amount paid by the next-largest component.

When considering the baseline costs shown in Figures 4.3 and 4.4, it is tempting to conclude that these are the costs associated with paying prorated S&I pay to drilling reservists who typically drill one weekend per month and perform AT for two weeks per year. Our data suggest that RC members receiving these pays serve more periods than what is typically assumed. Using our analytic file, we computed how many periods of ser-

FIGURE 4.2

Average Monthly Number of Reserve Component Members with AvIP Payment, by Component

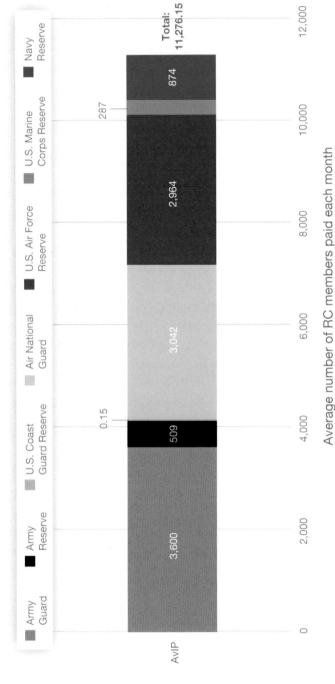

SOURCE: DMDC data, January 2018 to March 2020.

44

FIGURE 4.3

Annual Costs of HDIP Under Current Policy, by Pay and Component

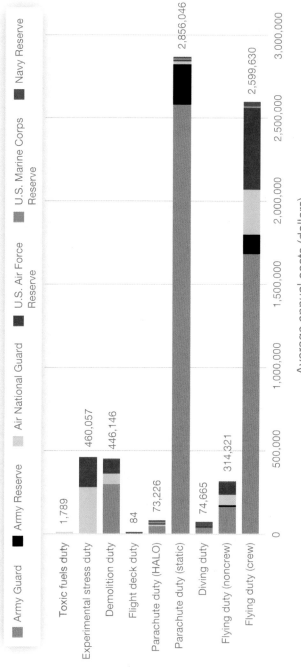

SOURCE: DMDC data, January 2018 to March 2020.

NOTE: Component-specific costs are listed in Appendix D. The U.S. Coast Guard Reserve had no reported payments of any HDIP during the observation window.

FIGURE 4.4

Annual Costs of AvIP Under Current Policy, by Component

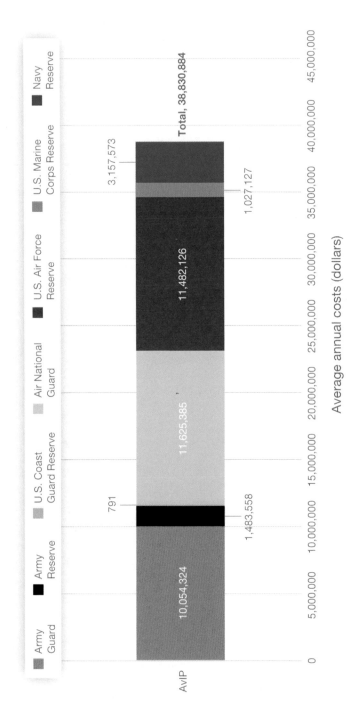

Average annual costs (dollars)

SOURCE: DMDC data, January 2018 to March 2020.

vice are compensated each time an S&I payment is made. Table 4.1 shows the average pay amount and the average number of periods being compensated each time a pay is disbursed. For comparison, the table also shows the expected average amount of one month's worth of each special pay, given the rates of pay shown in Chapter Two and under the assumption that RC members perform 14 days of AT and 48 IDT (drill) periods each year, with the member typically performing four drills each month.

The table shows that the average amount of each payment is greater than the expected amount. For example, the expected monthly amount for HDIP for flying duty (crew) is $35.17 under a strict schedule of 48 IDT drills and 14 AT days per year, but the average actual disbursement of this pay is $80. We find that the average amount of special pay is typically compensating for at least ten periods of service. For example, for HDIP for flying duty (crew), RC members on average are compensated for 11.8 periods each month they receive the pay. This is larger than the 5.17 periods we expected if RC members were performing only four IDT drills each month and 14 days of additional training. There are several explanations for this pattern. First, disbursements may be made infrequently but compensate for several months of service at once. This appears to be particularly true of experimental stress duty and toxic fuels duty pays, for which the average payment is compensating for more periods than can be served in a month. Second, RC members might serve more than the expected minimum each month. We investigate these explanations in more detail at the end of this chapter.

An important implication of the finding that drilling reservists are paid more on average each month than what we would expect is that the difference between what RC members get paid and the full rate will be smaller than expected. Put differently, special pay amounts under the current system are closer to the monthly rate than would be expected, given the minimum monthly service requirements of drilling reservists. Consequently, the difference in cost between the baseline policy and the full-rate policy will be smaller than what would be expected if drilling RC members served the assumed typical amount of duty time.

TABLE 4.1
Average Pay Amount and Periods Paid per Disbursement, by Pay Type

	HDIPs									
	Flying Duty (Crew)	Flying Duty (Noncrew)	Diving Duty	Parachute Duty (Static)	Parachute Duty (HALO)	Flight Deck Duty	Demolition Duty	Experimental Stress Duty	Toxic Fuels Duty	AvIP
Expected amount under "typical" drilling schedule	$35.17	$25.83	$36.39	$25.83	$36.25	$25.83	$25.83	$25.83	$25.83	$112.19
Average amount of each payment	$80	$61	$192	$84	$144	$95	$86	$581	$268	$287
Average number of periods paid in each payment	11.8	12.2	27.3	16.8	19.3	19.0	17.2	116.2	53.7	13.2

SOURCE: DMDC data, January 2018 to March 2020.

NOTE: A "typical" drilling schedule is assumed to be four drill periods of IDT each month and 14 days of AT each year, for an average of 5.17 periods served per month.

Counterfactual Estimates for Flying Hours

Using the tabulations provided by each RC, we calculated the percentage of all special pay earners who flew zero to 23 hours, 24 to 47 hours, or at least 48 hours in a year. If RC and AC flying hours requirements were made equal, RC members would need to fly 48 hours over the course of a year to maintain their eligibility for a full year.[1] Table 4.2 shows the results. The table suggests that few RC members who earn AvIP or HDIP for flying duty would lose eligibility if they were required to meet AC guidelines.[2]

Among those who earn AvIP, most fly at least 48 hours per year and so would not lose eligibility if the monthly requirements were raised from two to four hours per month. Some fly less than 23 hours per year, meaning they do not even reach current monthly requirements and so are likely on

TABLE 4.2
Percentage of AvIP Earners with Given Flight Hours in a Year

	AvIP		
	0–23 Hours per Year	24–47 Hours per Year	48 or More Hours per Year
Army Reserve	31%	4%	65%
Army Guard	21%	5%	74%
Air Force Reserve	53%	5%	42%
Air Guard	15%	6%	79%
Marine Corps Reserve	11%	6%	83%
Navy Reserve	0%	0%	100%

SOURCE: Data provided by RCs. We did not request data from the Coast Guard Reserve.

---

[1] For feasibility reasons, we requested aggregated annual information from the RCs. To maintain eligibility, an RC member's annual hours would have to be served in increments over multiple months, per the requirements set forth in Office of the Under Secretary of Defense (Comptroller), 2020, Ch. 22.

[2] Note that this is a hypothetical change; there is no indication in the FY 2020 NDAA that the flying hour requirements for receipt of AvIP or HDIP for flying duty will be revised.

continuous AvIP.[3] Around 5 percent fly between 24 and 47 hours each year. Some of these members could lose eligibility under a four-hour-per-month policy, but some are likely earning continuous AvIP, so their eligibility is not contingent on flying hours. According to this evidence, we believe a minority of AvIP earners would lose eligibility if flying hour requirements were to change. It is unclear what effect, if any, the difference in flying hours eligibility would have on costs, since some RC members could serve extra periods to make up the two-hour difference in flying hours each month. Therefore, the net effect on costs is uncertain but is not large compared with the overall costs of AvIP. We ignore the flying hours eligibility requirement in the rest of our calculations.

Full Cost Estimates

We estimated the costs of implementing special pays under a counterfactual policy that would pay a full monthly rate for each pay in every month that an RC member fulfills the duty eligibility requirements. We used the three methods described in Chapter Three.

Figure 4.5 shows the estimated annual increase in cost for HDIPs relative to the baseline policy. Aggregate annual cost increases for the pays shown here are between $7.1 million and $14.9 million. Most of these increases are attributable to flying duty for crew members and parachute duty for static jumps. Depending on the estimation method, total payments of flying duty for crew members would increase between $4.0 and $7.0 million per year. Parachute duty for static jumps would increase $1.5 million to $5.8 million per year. All other pays have expected increases of less than $1 million per year. The estimates from Method 2 for experimental stress duty and toxic fuels duty and Method 1 for experimental stress duty are not shown because they implied a decrease in costs.

---

[3] It is possible that those who fly fewer than 23 hours per year are not on continuous AvIP, but rather earn AvIP infrequently. However, in our data, RC members who earn AvIP at least once tend to earn it nearly every month. Therefore, it would appear that RC members who are not flying enough hours are likely on continuous AvIP.

FIGURE 4.5

Annual Cost Increases for HDIP Under Full-Rate Policy, by Pay Type and Estimation Method

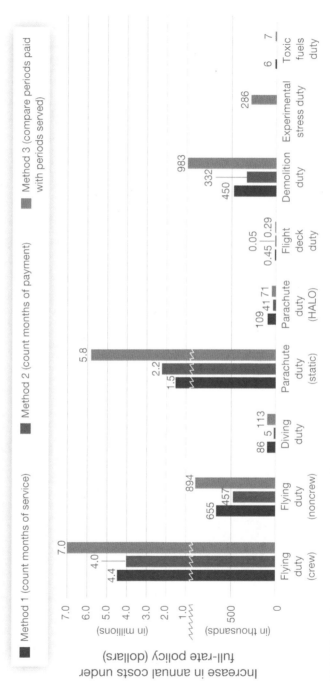

SOURCE: DMDC data, January 2018 to March 2020.

NOTE: Bars and labels are not shown when a method yielded a decrease for a particular pay.

Figure 4.6 shows estimated cost increases for AvIP under a full-rate policy. The increase varies from $39.1 million per year to $73.4 million per year. In the same way that the current costs for AvIP are far greater than the aggregate costs of HDIPs, these increases are much greater than the expected increase for HDIPs.

Figure 4.7 shows how these estimated increases translate to percentage increases above current costs. Aggregating across all pays, we estimate costs will increase between 90 percent and 150 percent. In terms of individual pays, the increase varies from around 50 percent to just over 200 percent.

Sensitivity Analyses

We conducted sensitivity analyses to determine how our cost analyses depended on assumptions we made while constructing our data set. We examined how much of our cost estimates are due to drilling reservists versus other RC members (who might train less frequently). We also examined whether our cost estimates potentially were influenced by payments

FIGURE 4.6

Annual Cost Increases for AvIP Under Full-Rate Policy, by Estimation Method

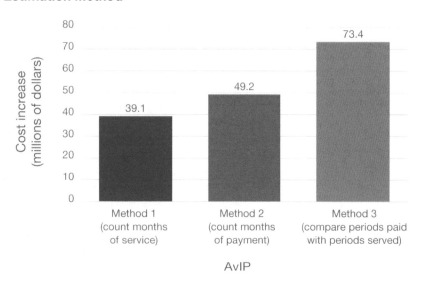

SOURCE: DMDC data, January 2018 to March 2020.

FIGURE 4.7

Percentage Increase in Annual Special Pay Costs Under Full-Rate Policy, by Estimation Method

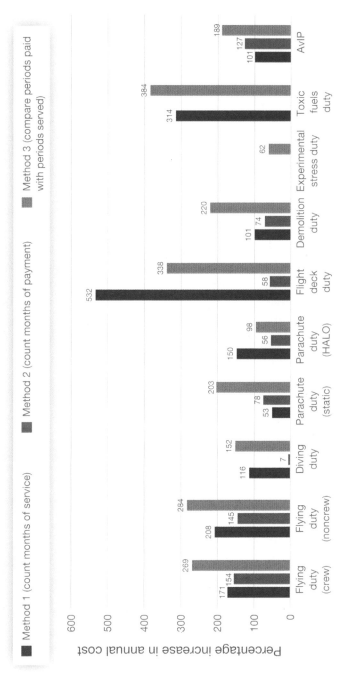

SOURCE: DMDC data, January 2018 to March 2020.

NOTE: Bars and labels are not shown when a method yielded a decrease for a particular pay.

from deployment that could be disbursed at a later date or by *clawbacks* (i.e., negative pay amounts) made in a subsequent month. Table 4.3 summarizes the results of these sensitivity checks.

First, we limited the sample to drilling reservists. Drilling reservists account for 94 percent of all HDIP and AvIP payments, ranging from 100 percent of flight deck duty to 92 percent of AvIP. They also account for 94 percent of total costs under the current system, with the smallest share being 89 percent of static-line parachute duty pay. When limiting our cost calculations to drilling reservists, we find that the expected increase in total costs changes 2.7 percent to 4.6 percent, compared with the increases shown in Figures 4.5 and 4.6.

Next, we included clawbacks in our estimates. Clawbacks are uncommon, occurring in less than 0.01 percent of all months. A clawback can occur if, for example, AvIP is paid out with the expectation that an RC member will fulfill the flying hours requirements for a six-month window, but then the member ultimately does not perform enough flight hours. We calculated the number of months clawed back in the same way as the number of months paid for each positive payment, with the exception that there is no way to account for clawbacks when counting eligibility using months of service. Including clawbacks decreases our estimates by less than 1 percent.

TABLE 4.3

Change in Cost Estimates Under Alternative Sample Specifications

Specification	Percentage Change Relative to Baseline		
	Method 1 (Service)	Method 2 (Payment)	Method 3 (Benchmark Pay to Service)
Limit sample to drilling reservists (RCC/TRC code "SA")	–2.8%	–2.7%	–4.6%
Include clawbacks in estimate	N/A	–0.7%	–0.8%
Exclude months after a deployment ends	N/A	–1.0%	–1.0%

SOURCE: DMDC data, January 2018 to March 2020.

NOTE: N/A = not applicable. Table shows percentage change relative to the aggregate cost increases shown in Figures 4.5 and 4.6.

Finally, we excluded months after a deployment has ended. Because payment does not align perfectly with service, payment for a deployment (including special pays) might be disbursed after the deployment has ended. This sensitivity check gauges whether special pays that appear to be earned during training might instead have been earned while deployed. We perform this sensitivity check only for the two methods that use pay data, since we cannot tie those payments to a precise month of service. Excluding months after a deployment decreases our estimates by about 1 percent.

Explaining the Magnitude of Cost Increases

To put these increases in perspective, consider what would be expected from the stereotypical training pattern of a drilling reservist. That is, suppose a reservist served exactly 14 days of AT in one month each year and exactly four IDT (weekend drills) in each calendar month. Further, suppose that they perform the hazardous duty every month and that the special pay is disbursed immediately, every month. Their pay amount would equal either $(14 + 4)/30$ (for the month with both AT and IDT service) or $4/30$ (for IDT-only months) of the full rate. This would average to 5.17 periods per month; paying the full rate would increase costs by a factor of 5.81.

None of our estimation methods suggest cost increases would be close to such orders of magnitude. There are several explanations for this, which we investigate in turn.

RC Members Serve More Than the Minimum Number of Periods Each Month

RC members' service patterns show that four drill periods of IDT do not constitute a "typical" month of service. Instead, RC members serve various combinations of duty statuses, if they perform service at all. Figure 4.8 shows how monthly service breaks down in our sample. Drilling reservists, who account for about 90 percent of the months in our sample, serve in more than one type of duty status in about 3.5 percent of all months: either ADT and IDT or a training status plus another type of active duty status. Moreover, in 34 percent of all months, they do not perform any service. In other words, instead of spreading service evenly over the course of a year,

FIGURE 4.8
Type of Duty Status Served in a Month, by RCC/TRC Code

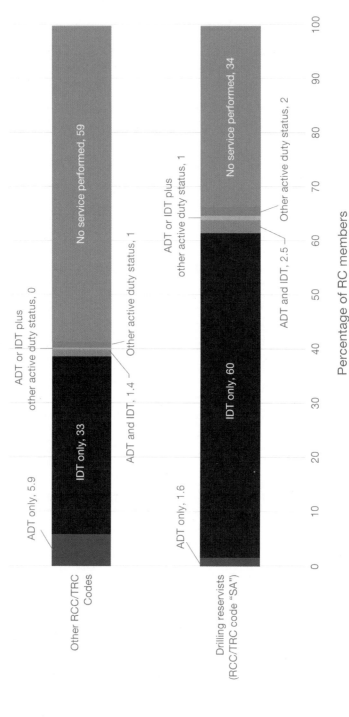

SOURCE: DMDC data, January 2018 to March 2020.
NOTE: Percentage totals do not sum to 100 because of rounding.

the average drilling reservist performs their training over the year in about eight months. The figure also shows that other RC members, who are not required to drill regularly, still train in 41 percent of months.

In addition to performing service in fewer than 12 months per year, RC members serve more than the minimum number of periods per month when they do serve. The average number of periods served in months with any training duty (i.e., the months represented by the dark blue, red, gray, and yellow bars of Figure 4.8) is 6.1. This is larger than the 5.2 expected of the stereotypical drilling patterns described earlier. It means that RC members who train end up serving more than one additional drill period, or an additional day of AT, or (perhaps) an additional day in another active duty status. Special pays would be prorated against the 6.1 period baseline, meaning that the full monthly rate would be a 4.9-fold increase, not the 5.8-fold increase predicted in the previous section. Therefore, the drilling patterns of RC members partly explain the magnitude of our cost predictions.

Pays Appear to Be Compensating for Multiple Months of Service

Although RC members serve more than the minimum required time each period, special pay disbursements appear to be covering even more than the actual service. Table 4.1 showed that the average disbursement of special pay compensates for more than ten periods of service. For instance, AvIP payments on average are worth 13.2 periods of service. Therefore, if we count one payment as one month of eligibility, a full-rate policy would just more than double the costs of AvIP, which is much lower than even the 4.9-fold increase expected from the actual observed amount of monthly service.

On the other hand, each disbursement may retroactively compensate for more than one month of hazardous duty. Our Method 3, which benchmarks periods paid to periods served, inflates or deflates each pay disbursement according to how it compares with observed patterns of service. This method is meant to account for the "lumpiness" of pays over time. In several cases, this method suggests an increase of roughly 280 percent to 380 percent for certain pays under a full-rate policy (or a 3.8- to 4.8-fold increase), as Figure 4.7 shows. This more closely approximates the expected increase that is based only on RC members' service patterns, suggesting that pay disbursements are somewhat unevenly distributed relative to service.

But there are still many unknowns. Appendix E provides a more detailed discussion of additional patterns in the data that hinder us from a full validation of our cost estimates. Further, as we discuss in Chapter Six, changes to the pay policy could induce behavioral changes that further increase the costs beyond the estimates presented here.

Summary

In this chapter, we presented three alternative approaches for estimating the increase in cost associated with paying RC members the full rate of S&I pays. These approaches were needed because of shortcomings in the data for our purposes. Each approach used a different method of estimating the number of months for which a member would be eligible to be paid the full rate of S&I pays. As a result, we have a range of possible cost increases, from $46.3 million to $88.5 million, or increases ranging from 100 percent to 194 percent. These cost increases are lower than expected. In part, this is due to RC members serving for fewer months than expected (eight, on average) and RC members serving more days or drills during a month than expected (6.1, on average).

CHAPTER FIVE

Potential Eligibility of Reserve Component Members for the Full Rate of Special and Incentive Pays

We calculated eligibility using the two methods described in Chapter Three. First, we counted the number of RC members who earn each pay in an average month. Next, we identified eligible occupations according to the percentage of those earning each pay who were assigned to a duty occupation. We calculated how many RC members were assigned to those eligible occupations each month to get our second estimate of the total eligible population.

For the second method, we chose an eligibility threshold of 5 percent: An occupation must account for 5 percent of all earners of a pay to be considered eligible for that pay. This threshold struck a balance between accounting for the majority of all RC members who earned pays versus having too many occupations included in our calculations. We calculate occupation-based eligibility for an RC only if it disburses a pay at least once each month.

Table 5.1 shows the eligibility results for each pay and component for the first and second methods. The actual earners per month are the same as those shown in Figure 4.1. The number of RC members in eligible occupation categories outnumber the actual earners, generally by one to two orders of magnitude. The difference shows that although a duty occupation may require the performance of a hazardous duty, most RC members in that occupation do not actually perform the duty.

Table 3.6 provided an example of how most Army Reserve members who earn flying duty (crew) pay come from aviation or medical occupations. But only a minority of those assigned to such occupations in any given month actually complete flight hours.

TABLE 5.1
Monthly Number of Eligible Reserve Component Members, by Pay Type and Component

Pay	Eligibility Calculation Methodology	Army National Guard	Army Reserve	Air National Guard	Air Force Reserve	Marine Corps Reserve	Navy Reserve
Flying duty (crew)	Actual earners per month	1,793	94	268	502	10	28
	Number of RC members in eligible occupation categories each month	29,871	15,297	3,091	4,158	96	425
Flying duty (noncrew)	Actual earners per month	215	6	105	91	<1	9
	Number of RC members in eligible occupation categories each month	30,718	9,727	14,774	17,943	—	3,194
Diving duty	Actual earners per month	15	0	2	1	1	13
	Number of RC members in eligible occupation categories each month	1,834	0	10,262	511	395	228
Parachute duty (static)	Actual earners per month	2,541	256	7	4	11	3
	Number of RC members in eligible occupation categories each month	123,072	28,471	11,337	211	1,809	153

Table 5.1—Continued

Pay	Eligibility Calculation Methodology	Army National Guard	Army Reserve	Air National Guard	Air Force Reserve	Marine Corps Reserve	Navy Reserve
Parachute duty (HALO)	Actual earners per month	36	<1	1	3	1	1
	Number of RC members in eligible occupation categories each month	1,834	—	11,052	3,395	225	175
Demolition duty	Actual earners per month	269	<1	63	93	0	7
	Number of RC members in eligible occupation categories each month	3,518	—	21,858	4,813	0	175
Experimental stress duty	Actual earners per month	<1	0	39	27	0	0
	Number of RC members in eligible occupation categories each month	—	0	3,518	511	0	0
AvIP	Actual earners per month	3,600	509	3,042	2,964	287	874
	Number of RC members in eligible occupation categories each month	4,426	803	6,569	6,397	696	1,606

SOURCE: DMDC data.

NOTE: Pays not shown had an average of less than one RC member among all components who earned the pay each month. The U.S. Coast Guard is not shown; there were zero reported RC members earning HDIP in the U.S. Coast Guard Reserve and less than one per month earning AvIP.

CHAPTER SIX

Paying Reserve Component and Active Component Members the Same: Discussion of Broader Issues

The request from Congress for a report on paying RC members the full rate of S&I pay for hazardous duty and AvIP stemmed from concern about the equitable payment of S&I pay for RC members and AC members. In this chapter, we review the arguments for and against prorating S&I pay using the one-thirtieth rule, particularly for hazardous duty pay and AvIP, drawing on analysis and discussion from the sixth, ninth, and eleventh Quadrennial Review of Military Compensation (QRMC). We then discuss the readiness implications of paying the full rate rather than the prorated amount of S&I pay, drawing on evidence provided by RAND's dynamic retention model (DRM). Finally, we place the discussion of equity with regard to S&I pay within the broader context of equity of compensation for RC and AC members, highlighting other ways in which compensation differs for each group and ways in which AC and RC service differs.

Conclusions from the Sixth, Ninth, and Eleventh Quadrennial Review of Military Compensation

The reports of the sixth QRMC in 1988, the ninth QRMC in 2002, and the eleventh QRMC in 2012 all considered the advisability of prorating S&I pay for members of the RC (DoD, 1988, 2002, and 2012). The sixth QRMC supported the continued use of the one-thirtieth rule to prorate S&I pays for RC members, stating,

Reservists on ADT or IDT performing certain specialties receive 1/30th of the basic pay rate for active duty members for each period of duty performed. The 6th QRMC believes this rate of pay is both appropriate and consistent with the manner in which members of the reserve components are compensated (DoD, 1988, pp. 5–7).

In contrast, the ninth QRMC concluded that further study was necessary to determine the appropriateness of prorating RC S&I pays and that it would be fundamental for such a study to consider the changing role of the reserves and whether changing S&I pay policy would have a positive impact on recruiting and retention.[1] That said, the ninth QRMC also stated that there are some S&I pays (specifically, those based on exposure to hazard or hardship) for which a more consistent application of S&I pay for RC and AC members might be considered. As described by the ninth QRMC, hazardous duty and hardship duty pays are designed either to encourage participation in specific onerous or dangerous duties or to reward members who encounter danger while performing an assigned duty. Both pays help ensure the retention of personnel who perform these duties or face these dangers. AC members must reach a certain threshold in terms of exposure to hazards or hardships to be entitled to the full pay.[2] But RC members who meet or exceed the same threshold as AC members are paid at the one-thirtieth rate, which is inconsistent. The exception is HDIP for flying duty, for which the RC threshold is lower. The ninth QRMC said it could be argued that for cases in which RC members meet or exceed the same threshold as AC members, they should receive more than the one-thirtieth payment currently allowed. On the other hand, it also acknowledged the counterargument that AC members, on average, have greater exposure to hazards, given their full-time status. Furthermore, AC members who exceed the thresholds are paid the same (full rate) as other AC members who just meet the threshold. Thus, the ninth QRMC concluded that the policy was inconsistently applied for

---

[1]   The ninth QRMC drew on analysis by Hogan et al., 2002.

[2]   The exception is HFP, for which both AC and RC members receive the full monthly payment if they spend even one day in the specified zone during the month, regardless of their duty status. This pay is not applicable to IDT.

both RC and AC members for S&I pays related to hazardous or hardship duty.

Regarding AvIP, the ninth QRMC considered this pay a "Career and Skill Pay with Hazardous Duty" that provides an incentive to pursue a career in aviation while providing partial compensation for the expectation of hazardous duty. To the extent that the use of the RC has expanded and reserve careers in aviation need to be encouraged, paying one-thirtieth of the AC rate may not be enough to ensure a sufficient supply of RC aviators. However, the ninth QRMC stated that more-detailed study was needed to ascertain whether this was the case.

The eleventh QRMC argued that whether the one-thirtieth rule should be eliminated depends on the effect of such a policy change on readiness or operational capability. If receiving the full monthly S&I pay increased the participation of RC members, then it could be presumed that readiness and/or operational capability increased as well. But if participation does not increase, then the only outcome of adopting a full-rate policy would be to increase personnel cost. The eleventh QRMC acknowledged that prorating monthly incentive pays has been an ongoing issue for some RC members and that some Coast Guard and reserve members have argued that the one-thirtieth rule for setting S&I pays is unfair, given that they typically must perform the same amount of duty as AC members to be eligible for the pay and, in some cases, perform more than the eligibility threshold.

In the case of hazardous duty pay, however, the eleventh QRMC argued that the military cannot put a member in a situation in which they are exposed to a relevant hazard duty unless the member is "on duty"; because RC members who perform AT and IDT are only on duty part time, it would not be appropriate to compensate RC members for hazards that they could not be exposed to. When RC members are activated and on duty full-time, their hazardous duty pay is paid at the full rate, the same as for AC members. In the case of AvIP, the eleventh QRMC stated that this S&I pay is a career incentive pay. RC members choose to serve less than full-time, and the one-thirtieth rule reflects this career decision: It not only compensates RC members in proportion to their participation but also provides a positive incentive for RC members to increase participation efficiently and effectively. More broadly, the eleventh QRMC expressed concern that most RC members would not increase participation if compensation did not increase in proportion to participation, thereby increasing cost without improving operational capability. The eleventh QRMC concluded that

> while equity is a relevant consideration in evaluating elements of a military pay structure, the effectiveness and efficiency of the pay system in maintaining readiness and operational capability is an important concern as well and will take priority in many circumstances (DoD, 2012, p. 177).

Assessing the Effects of Eliminating Proration on Readiness

The QRMC reports argue that any change to the one-thirtieth rule for setting S&I pay should be based on force management and readiness considerations. As mentioned, the eleventh QRMC concludes that, without proration of S&I pay and the increase in compensation associated with participation, most reservists will not change their level of participation. In this section, we explore the force management effects of eliminating the proration of S&I pays and paying the full rate of S&I pay to RC members. Specifically, we consider how eliminating proration would affect reserve participation, drawing on insights from the economics literature and by making use of RAND's DRM.

Paying members of the RC the full rate of S&I pays can affect the level of RC participation in two ways. The first way—what economists call the *intensive margin*—is the effect on the intensity of RC participation, or the number of drills and days per month that a member would choose, among Selected Reservists already in the RC. The second way—what economists call the *extensive margin*—is the effect on the level of RC membership: that is, the number of people who decide to be in the Selected Reserve.

To consider these effects theoretically, we extended the classic journal article in the economics literature on multiple job holding by Robert Shishko and Bernard Rostker, 1976.[3] In the model, an individual maximizes utility by choosing the amount of time to spend working at a secondary

---

[3] Our extension is presented in Appendix B.

job, given that they can only work a fixed amount of time at their primary job. Although the model can be applied to any secondary job decision, it has often been used to conceptually model decisions about the intensity of participation in the Selected Reserve: i.e., the number of drills and days an individual spends on reserve duty in a month (Gilroy, Horne, and Smith, 1991; Kocher and Thomas, 1990; Mehay, 1991). With respect to the intensive margin, our extension demonstrates that prorating S&I pay provides the same incentive effect on the intensity of RC participation in a month as basic pay does, which is also prorated. If increasing basic pay increases the intensity of RC participation, then so does increasing prorated S&I pay. However, if proration of S&I pay were eliminated and RC members who met the minimum threshold were paid the same (full) rate of S&I pay, regardless of the intensity of their participation, our extension demonstrates that the incentive effect of S&I pay on the intensity of participation in a month is unambiguously negative if RC members prefer to work less when their unearned income rises. Thus, economic theory implies that the intensity of RC participation would decrease under a full-rate S&I pay policy.

With respect to the extensive margin, we show in Appendix B that economic theory implies that the full-rate policy would induce more people to decide to be in the RC (though the intensity of their participation in a month would be less). We further explore this implication using the DRM for Air Force pilots. The Air Force pilot community is useful to consider when examining the impact of a change in S&I pay policy. The Air Force uses only two S&I pays to manage the retention of pilots and to address external market forces that can affect the retention of military aviators: specifically, AvIP and Aviation Bonus (AvB). Unlike AvIP, Air Force pilots in the AC receive AvB only if they commit to a multiyear obligation that typically varies with the specific rated occupation and the length of the obligation incurred.[4]

---

[4] Three common options that have been offered by the Air Force are a three-year contract, a five-year contract, and a contract until 20 YAS at amounts that are now up to $35,000 per year and previously were up to $25,000 per year. The portfolio of AvB contracts offered has changed over time, and a history of AvB is provided in Mattock et al., 2016. We note that both AC and RC members can be offered an AvB, as provided in 37 U.S.C. § 334(b), as a force management tool available to the services.

RAND researchers estimated a DRM of Air Force pilot AC retention and RC membership that can be used to explore the effect of alternative RC S&I pay policies (Mattock et al., 2016). The DRM is an econometric model that shows how AC retention and the level of RC membership are affected by changes to pay and personnel policies. Although the model accounts for the level of RC membership, it does not account for the intensity of RC partici-pation (i.e., the number of drills and days a member participates per month): All RC members are assumed to have 14 days of AT and 60 IDT and other drill periods per year for which they are eligible to receive prorated AvIP. We use the DRM for Air Force rated personnel to simulate the effect of paying RC members the full rate of AvIP.

Figure 6.1 shows the simulated steady state retention profile for Air Force rated personnel in the baseline versus the alternative policy. AC retention is on the left, and RC membership is on the right. The baseline (the gray line) is the predicted retention or membership at each year of service under cur-rent Air Force AvB and AvIP policy, in which AvIP is prorated. The red line is the predicted retention if RC members were to be paid the full rate of AvIP instead.

The simulation results show that eliminating the proration of AvIP and paying the full rate to RC members each month would reduce AC retention and increase RC membership. The increase in RC membership is consis-tent with the theory regarding the extensive margin, which we explore in Appendix A. The increase in S&I pay caused by changing the policy from proration to the payment of the full rate induces more people to join and participate in the RC. The DRM results indicate that the overall change in RC membership would be substantial (7.3 percent), and the change is more or less evenly split between those with less than 20 years of combined AC and RC service and those with 20 or more years of combined service. More surprising is the decline in AC retention, although the overall decline in AC retention is relatively small (1.1 percent), with the decline among those with less than 20 years of service equal to 0.8 percent and among those with 20 or more years of service equal to 2.8 percent. The decline occurs because Air Force pilots find serving in the RC relatively more attractive than serving in the AC when AvIP is paid at the full rate to RC members.

The changes in the experience mix of AC and RC personnel caused by the changes to AC retention and RC membership shown in Figure 6.1 imply

FIGURE 6.1
Air Force Pilot Retention

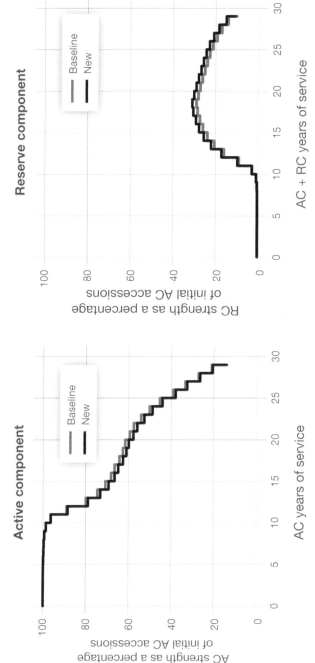

Active component

AC strength as a percentage
of initial AC accessions

AC years of service

Change in force = –1.1%; change before 20 years of
service = –0.8%; change after 20 years of service = –2.8%

Reserve component

RC strength as a percentage
of initial AC accessions

AC + RC years of service

Change in force = 7.3%; change before 20 years of
service = 7.3%; change after 20 years of service = 7.2%

Baseline
New

SOURCE: DMDC data, January 2018 to March 2020.

that personnel costs would change. The DRM also allows us to compute per capita personnel cost changes, for which personnel costs include the cost of basic pay and allowances, AvB, AvIP, and retirement accrual costs. Table 6.1 shows that per capita cost falls for AC Air Force rated personnel by just less than $600; this is largely driven by a decline in seniority in the AC. The per capita cost for the RC rises by nearly $8,000 because of the change to paying RC members the full rate of AvIP regardless of the number of days or drills they serve in a month, provided they qualify for AvIP.

In response to an increase in RC compensation, the AC might reasonably be expected to increase the AvB to sustain retention and reduce the separation of pilots to the RC. In our simulation, for example, the Air Force could restore AC retention by raising the AvB from $35,000 per year to $38,250.[5] This would result in the per capita AC cost increasing to $132,954, or about $1,200 over the baseline. This increase in AC compensation would diminish the effect of the increase in RC S&I pay, reducing the increase in RC membership by 2.1 percentage points, from 7.3 percent to 5.2 percent. Thus, the need to meet staffing requirements for the higher pay grades and more-experienced aviators could actually drive personnel costs up rather than reduce them. As a result, a reduction in the AC per capita cost is unlikely to be realized if proration of S&I pay is eliminated.

In sum, our analysis suggests that eliminating the proration of S&I pay would reduce the intensity of RC participation but increase the number of personnel who seek membership, provided there are either unfilled RC billets for aviation personnel or the number of billets is increased to accom-

TABLE 6.1

Per Capita Cost of Regular Military Compensation, Retirement, AvB, and AvIP Under Baseline and Full-Rate Scenarios

	AC Per Capita Cost	RC Per Capita Cost
Baseline	$131,742	$88,410
Full-rate AvIP for RC	$131,165	$96,296

SOURCE: DMDC data, January 2018 to March 2020.

NOTE: Regular Military Compensation and retirement costs are inflation-adjusted to 2021 dollars.

---

[5] As of 2021, AvB is capped at $35,000.

modate the additional aviation personnel being accessed into the RC. Additional analysis is required to ascertain whether the increase in the number who participate would be sufficient to offset the intensity of participation by a given reservist. In fact, if there are no vacant billets or the number is not increased, then a decline in the intensity of RC participation would lead to a decline in readiness and the ability to meet mission requirements. The analysis also indicates that eliminating proration of S&I pay in the RC could reduce AC retention, other things (including AvB) held constant. In addition to the intensive and extensive margins we discuss here, we might also expect that eliminating proration could increase the incentive to game the system by strategically timing the performance of training (e.g., timing AT duty to span two months rather than one month, perhaps by starting training at the end of one month and continuing it in the next month, thereby resulting in the receipt of two months of full-rate S&I pay). Additional analysis is also required to ascertain the extent to which such gaming would occur.

CHAPTER SEVEN

Feasibility and Advisability of Paying Eligible Reserve Component Members the Full Rate of Special and Incentive Pays

The congressional requirement for this study included a mandate to assess the feasibility and advisability of paying eligible RC members full-rate S&I pays. To assess feasibility, we put our estimates from Chapters Four and Five of the costs and number of personnel who might be affected by a full-rate policy into a larger context in terms of the overall size and budgetary cost of RC personnel. We note that our assessment does not consider the administrative costs of implementing a full-rate RC policy.[1] To assess advisability, we consider the implications of a full-rate policy for readiness and efficiency, as discussed in Chapter Five and raised by past QRMCs. We conclude that paying RC members the full rate of S&I pays may be feasible in terms of cost, since that cost (excluding implementation costs) would be quite small relative to the size of the RC personnel budget. However, paying RC members the full rate of S&I pays might not be advisable because it would be inefficient and could adversely affect readiness.

---

[1] Our analysis also ignores the costs of paying active duty members the full rate of these S&I pays. Such an analysis was not required by the NDAA, but removing the proration of these pays would also affect active duty members who qualify for less than a full month. We do not know how many such active duty members there are, or how much such a policy change would cost.

Feasibility

Our estimates from Chapter Four show that the increase in cost associated with paying HDIP at the full rate for RC members would be (at most) $88.5 million annually, 194 percent over the baseline of $45.7 million annually. From the point of view of the baseline S&I pay budget, this is a substantial increase. On the other hand, it is small relative to the overall Reserve and Guard personnel budgets (less than 0.4 percent for FY 2021).[2] We note that these cost estimates assume that individual RC member participation would be unchanged under a full-rate policy, although (as we discuss in the context of advisability) members would have an incentive to change their participation behavior.

In terms of the number of personnel eligible for a full-rate S&I pay, the number of personnel who would be affected by a full-rate policy for HDIP would be as high as 0.35 percent for parachute duty (static) when eligibility is measured in terms of the number of RC members who earn the pay each month on average, assuming an overall RC force size of 800,349 as of March 2020.[3] That is, when we measure eligibility in terms of the number of personnel who perform duties that result in HDIP payment, the share of personnel is, at most, less than 1 percent of the RC force. On the other hand, if we measure eligibility in terms of the potential number of personnel who could be eligible by counting the number of personnel in occupations that have at least 5 percent of RC members receiving the pay, the share of members eligible increases to 21 percent (again for parachute duty [static]). As we discussed in Chapter Five, the difference in these estimates shows that although a duty occupation might require the performance of a hazardous duty, most RC members in that occupation do not actually perform the duty.

In sum, these results indicate that the increase in costs would be relatively small as a share of the RC personnel budget, though the percentage increase relative to the baseline S&I pay budget would be large. Furthermore, the share of personnel who earn these pays relative to the overall size of the RC is less than 1 percent, but the share of personnel who could earn

---

[2] See Office of the Under Secretary of Defense (Comptroller), 2021.

[3] See DMDC, undated.

a full-rate S&I pay could be as high as 20 percent. Again, it is important to note that these estimates do not include any changes in cost associated with changes in RC participation behavior caused by a full-rate policy.

Advisability

Paying RC members the full monthly rate of S&I pays might not be advisable from the standpoint of readiness and efficiency. As shown in the previous chapter, the full-rate policy would give an incentive to members to serve less intensely in any given month. That is, incentives to participate for more than the minimum number of days or drills required in a month would fall under a full-rate policy. After all, from the perspective of pay, why would an RC member participate more than the minimum under the full-rate policy when the amount of S&I pay remains unchanged? With proration, by contrast, the amount of S&I pay increases with the amount of participation, up to the full rate of the S&I pay. This is not to say that money is the only reason why members participate in the RC. Some members might continue to participate more than the minimum even under the full-rate policy. Instead, the analysis indicates that, for RC members at the margin of deciding whether to participate more, the full-rate policy gives a disincentive to participate more than the minimum.

The reduced incentive to participate more than the minimum could be important because our analysis indicates that drilling reservists serve more training periods, on average, than the stereotypical four IDT drills per month and 14 days of AT per year. That is, because drilling RC members serve more than the minimum amount of service on average, a full-rate system could reduce readiness by reducing the average number of drills or days per RC member served per year. By implication, RC members would tend to accumulate less experience in their occupations.

The reduced intensity of participation would mean that a full-rate policy would not only reduce readiness but also be inefficient. The RC would experience an increase in S&I cost, as summarized in the previous section; at the same time, RC members in any given month would have a reduced incentive to participate beyond the minimum and, by implication, would accumulate less experience. Thus, the RC would be less ready but at higher cost.

Offsetting this effect is the stronger propensity of RC members to join the RC in those occupations and duties for which the full rate of S&I pay is paid. As shown in the previous chapter, the full-rate policy would increase the number of members who participate in the RC, even while reducing the intensity of participation in any given month. The implication is that although individual members might participate less in any given month and accumulate less experience, there would be more members overall available to meet the mission (provided that the total number of members does not exceed RC end strength constraints and there are billets available that require particular skills for which incentives are being paid). Therefore, the composition of the RC would change toward less-experienced members. Furthermore, we find that a full-rate policy would draw members away from the AC and toward the RC. Consequently, AC retention would fall, and these members would need to be replaced with new recruits to maintain AC strength, and/or AvBs would need to increase. The AC could become more junior as well. Decisionmakers would need to carefully consider the trade-off between having fewer AC and RC members with more experience and having more individuals at a lower average experience level.

One thing is clear: Paying RC members the full rate of S&I pays would be inefficient if the goal is to provide an incentive to participate for more than the minimum number of drills or days in a month. In fact, it would have the opposite effect: As long as individuals positively value leisure, increasing their wealth by the amount of the full rate of S&I pay each month would give RC members an incentive to reduce the time they devote to RC participation.

Reserve Component Pay Authorities and Duty Statuses

In this appendix, we discuss two factors that are necessary to understand the differences between AC and RC payment of S&I pay: (1) how AC and RC members are compensated through 37 U.S.C. § 204 and 206 and (2) the relevant types of reserve duty status.

The documentation of S&I pay amounts for AC and RC members typically relies on two authorities: 37 U.S.C. § 204 and 206. By and large, section 204 stipulates that AC members or RC members who are called to active duty receive basic pay, and section 206 stipulates that RC members are entitled to a proportion of basic pay. Following this discussion of pay authorities, we describe *duty statuses*: statutory authorities under which RC members are called to serve. Understanding duty statuses is important for understanding the monthly compensation RC members receive and, therefore, understanding monthly S&I pay compensation.

Pay Authorities

37 U.S.C. § 204

Section 204 states that service members on active duty or in several other cases receive basic pay according to years of service and pay grade. RC members qualify for basic pay under section 204 in three circumstances: (1) if they are called to active duty, including AT exercises; (2) if they are physically disabled during active duty or IDT; or (3) if they are performing funeral honors duty under 10 U.S.C. § 12503 or 32 U.S.C. § 115 (37 U.S.C. § 204). RC members who are training full-time, including annual two-week training exercises, are considered to be on active duty and thus receive basic pay under 37 U.S.C. § 204 (DoDI 1215.06, 2015). Service members who begin or end eligibility for basic pay under 37 U.S.C. § 204 receive a prorated amount of basic pay: one-thirtieth of basic pay for each day during the month they were eligible for basic pay (37 U.S.C. § 204).

37 U.S.C. § 206

Section 206 states that RC members who are not entitled to basic pay under 37 U.S.C. § 204 are eligible for compensation equal to a fraction of the basic pay that they would be entitled to according to their pay grade and years of service.[1] RC members receive compensation equal to one-thirtieth of basic pay for the corresponding pay grade and years of service for each period of regular instruction or appropriate duty of at least two hours. For example, if an RC member participates in four drills over one weekend in one month, they would be paid four-thirtieths of basic pay that month (37 U.S.C. § 206).

Reserve Duty Status

Unlike AC members, who always serve under one duty (active duty), RC members serve under a variety of duties, called duty statuses. There are duty statuses under which RC members are considered active duty (working full time for their service) or inactive duty (working part time for their service). Duty statuses are statutory authorities that lay out the situations in which RC members are called to serve. Duty statuses include weekend drills, medical treatment, funeral honors, and full mobilization to active duty; each duty status typically falls under one of the following two categories: active or inactive. Table A.1 shows the five primary duty types, whether they are considered inactive duty or active duty, a description of how their compensation would be calculated, and the relevant authorities.

---

[1] Section 206 uses the word *compensation* to describe the pay that RC members receive. This is somewhat misleading. *Compensation* typically refers to all monetary compensation that a service member receives, including basic pay, basic allowance for housing, and other special pays (if applicable). However, the compensation that 37 U.S.C. § 206 describes is a proportion of basic pay.

Duty statuses for RC members serving on inactive duty typically include IDT (most commonly, weekend drills), muster duty, and funeral honors duty. Duty statuses for RC members serving on active duty typically include annual two-week (or longer) training, initial training, and many other duty statuses, including full mobilization to active duty or medical treatment. Many (and likely most) RC members typically perform duty prescribed in 10 U.S.C. § 10147, which describes the following training requirements for RC members: 48 drills per year (typically four drills per weekend, one weekend per month) for IDT and a training period of at least 14 days serving on active duty each year for AT (10 U.S.C. § 10147). Instead of 10 U.S.C. § 10147, National Guard members typically perform their annual training requirements under 32 U.S.C. § 502(a). In this report, we are primarily concerned with IDT and AT.

Each duty status is associated with a different monthly pay structure for RC members, which we detail in the "compensation description" column in Table A.1. S&I pay for RC members on an inactive duty status is often prorated according to monthly compensation; thus, it is important to understand how RC members are compensated each month (DoDI 1215.06, 2015).

TABLE A.1
Reserve Component Duty Types, Descriptions, and Compensation Structure

RC Duty Type	Description	Compensation Description	Inactive Duty or Active Duty	Receive Basic Pay Under 37 U.S.C. § 204 or Compensation Under 37 U.S.C. § 206	Relevant Authorities
IDT	Typically monthly, weekend drills	48 drills per year are required. Individuals are compensated at the rate of one-thirtieth of basic pay for each drill or period of instruction.	Inactive	206	10 U.S.C. § 10147(a)(1)
Muster duty	Annual medical and personnel information screening and collection effort. Occurs once per year and must last at least two hours. RC members receive a per diem for their time.	Compensated at 125 percent of the average per diem in the United States.	Inactive	N/A	10 U.S.C. § 12319
Funeral honors duty	Perform funeral honors at a veteran's funeral.	Individuals are compensated through section 206 or 495.	Inactive	206 or 495 (eligibility for incentive pays is granted only if compensated under 206)	10 U.S.C. § 12503

Table A.1—Continued

RC Duty Type	Description	Compensation Description	Inactive Duty or Active Duty	Receive Basic Pay Under 37 U.S.C. § 204 or Compensation Under 37 U.S.C. § 206	Relevant Authorities
ADT	There are three categories: AT, a two-week or longer annual training; IADT, for those with no prior military service; and OTD, other full-time skills and refresher training.[a]	Individuals are compensated through section 204 and receive basic pay.	Active	204	AT: 10 U.S.C. § 10147(a)(2); IADT and OTD: 10 U.S.C. § 12301(d)
Active duty other than training	Includes any other time when an RC member is called to serve under active duty, including mobilizing for a war or national emergency or receiving medical care.	Individuals are compensated through section 204 and receive basic pay.	Active	204	There are many. Examples include 10 U.S.C. § 12301(a), 10 U.S.C. § 12302, and 10 U.S.C. § 12304.

SOURCE: DoDI 1215.06, 2015.

NOTE: OTD = other training duty.

[a] DoDI 1215.06, 2015, p. 11, classifies AT, IADT, and OTD all as ADT. AT, IADT, and OTD are all active duty statuses and are compensated under 37 U.S.C. § 204. In practice, AT is often thought of as separate from IADT and OTD because it usually happens more frequently (once per year) and for a shorter period (typically, about 14 to 15 days).

Modeling Reserve Component Participation Under Alternative Special Pay Policies

In this appendix, we extend the model presented in Shishko and Rostker, 1976, which explains labor force participation in secondary or "moonlighting" jobs. In the original model, an individual maximizes utility by choosing the amount of time to spend working at a secondary job, given a fixed amount of time at the primary job and the utility they derive from leisure time. In our application, we can think of the secondary job as the intensity of participation in the selected reserve: that is, the number of drills or days an individual spends on reserve duty in a month. We first present the original model in Shishko and Rostker, 1976, then propose an extension to examine the effect of moving from a policy in which individuals receive S&I pays that are prorated according to the number of drills or days of qualifying duty in a month to a policy in which individuals receive the full rate of S&I pay regardless of the number of drills or days in a month, so long as they serve the minimum qualifying time in that month (e.g., two hours per month or 12 hours in six months for AvIP).

Shishko and Rostker's Model of Labor Supply for a Secondary Job

In the model, an individual maximizes utility by choosing x, a consumption good, and l, the amount of leisure time they have available. The maximum amount of the consumption good they can afford is the sum of their nonlabor income I_0 and the total wages they earn at their primary job and from Reserve service, $w_p L_p + w_R L_R$. The maximum amount of leisure time they have is the total time available in a month minus the time spent on their primary and secondary jobs, or $N - L_p - L_R$. The strategy that Shishko and Rostker used to analyze the model is to take the first order conditions, totally differentiate the first order conditions to solve for various relationships of interest (such as how an increase in the wage for the secondary job affects the number of hours spent on the secondary job), and use the second order conditions to determine the sign of the relationship (positive, negative, or ambiguous). Note that we use the same equation numbering as in Shishko and Rostker, 1976, for ease of reference. Table B.1 lists the parameters.

TABLE B.1

Symbols Used in Mathematical Model

Parameter	Explanation
$U(x, l)$	Utility function, assumed to be twice continuously differentiable and quasiconcave
x	Consumption good, also numeraire
l	Leisure time
I_0	Nonlabor income
w_p	Wage of primary job
L_p	Monthly time at primary job
w_R	Wage from Reserve service
L_m	Monthly time spent in Reserve service
N	Total time available in a month
U_l	Partial derivative of U with respect to l
U_x	Partial derivative of U with respect to x

The constrained utility-maximization problem is as follows:

$$\max U(x, l)$$

s.t.

(1) $I_0 + w_p L_p + w_R L_R - x \geq 0$

(2) $N - L_p - L_R - l \geq 0$

(3) $x, L_R, l \geq 0.$

The first order conditions are

(7) $U_l + U_x w_R = 0$

(8) $I_0 + w_p L_p + w_R L_R - x = 0$

(9) $N - L_p - L_R - l = 0$

(10) $\Lambda \begin{vmatrix} dL_R \\ dl \\ dx \end{vmatrix} = J,$

where

$$\Lambda = \begin{vmatrix} w_R & 0 & -1 \\ -1 & -1 & 0 \\ 0 & (w_R U_{xl} - U_{ll}) & (w_R U_{xx} - U_{xl}) \end{vmatrix}$$

and

$$J = \begin{vmatrix} -dI_0 - w_p dL_p - L_p dw_p - L_R dw_R \\ dL_p \\ -U_x w_R \end{vmatrix}$$

$$|\Lambda| = -w_R (w_R U_{xx} - U_{xl}) + (w_R U_{xl} - U_{ll}) = \frac{|H^*|}{U_x^2} > 0,$$

where H^* is the bordered Hessian of the utility function, and the determinant of the bordered Hessian is positive at a maximum.

Using these calculations, Shishko and Rostker derive some relationships that will be useful to us in modeling the effect of changing S&I policy to pay

the full rate of S&I pay to members of the RC. The first relationship of interest is the effect of wage w_R on the level of intensity L_R:

$$(11) \quad \frac{\partial L_R}{\partial w_R} = \frac{U_x}{|\Lambda|} - L_R \frac{\Lambda_{11}}{|\Lambda|},$$

where Λ_{11} is the cofactor of the 1,1 element of Λ, $\Lambda_{11} = -w_R(w_R U_{xx} - U_{xI})$.

Shishko and Rostker note that this is the Hicks-Slutsky decomposition,
in which the first term, $U_x / |\Lambda|$, is interpreted as the opportunity cost of
not performing RC service, and the second term is the "income term." The
first term is unambiguously positive. In the second term, $\Lambda_{11} / |\Lambda|$ is positive
if leisure is a superior good (that is, if more leisure is consumed as wealth
increases), so the second term as a whole is negative. Thus, the sign on the
right-hand side of Equation 11 is ambiguous because it depends on the relative magnitudes of the first term and the second term.

The second relationship of interest to us is the relationship between labor
supply to the reserves, L_R, and nonlabor income, I_0:

$$(14) \quad \frac{\partial L_R}{\partial I_0} = -\frac{\Lambda_{11}}{|\Lambda|}.$$

The effect of nonlabor income on secondary occupation labor supply is
unambiguously negative if leisure is a superior good.

Extending the Shishko and Rostker Model to Examine Changes in Reserve Component Special and Incentive Pay Policy

In this section, we show a simple extension of the Shishko and Rostker
model that can demonstrate that the effect of paying S&I pays at the full
rate rather than prorating them to the number of drills or days served will
unambiguously reduce the amount of labor supplied by individuals in the
Selected Reserve as long as leisure is a superior good. (For the moment, we
leave aside the question of whether paying S&I pays at the full rate would
encourage more individuals to enter the Selected Reserve because of the
higher expected wage at the minimum level of participation.)

First, we redefine I_0 to be

$$I_0 = I_1 + \rho S,$$

where I_1 is nonlabor income, S is the full rate of S&I pay, $\rho = 1$ if the full-rate policy is in effect, and $\rho = 0$ if the proration policy is in effect. Thus, ρ is an indicator variable of the policy regime. Note that S is only paid every month provided that the RC member participates at the minimum level. Under current policy, $\rho = 0$, and the proportion of S that individuals receive depends on the number of drills or days they have in a given month.

Second, we redefine w_R to be

$$w_R = w_B + (1 - \rho)\frac{1}{N}S.$$

That is, w_R is decomposed into w_B (conceptually, "basic pay") and the pro-rated S&I pay,

$$\frac{1}{N}S,$$

where N is 30 under current policy. If $\rho = 0$, then this expression reflects the current policy in which S&I pay is prorated; if $\rho = 1$, this expression would reflect the policy in which RC members receive the full rate of S&I pay.

Given this structure, we can look at the effect of changing S on labor supply L_R and how the policy regime $\rho = 1$ versus $\rho = 0$ affects labor supply. First, we find the partial derivatives of w_R and I_0 with respect to S, then we use the chain rule to find

$$\frac{\partial L_R}{\partial S}:$$

$$\frac{\partial w_R}{\partial S} = (1 - \rho)\frac{1}{N}$$

$$\frac{\partial I_0}{\partial S} = \rho.$$

The change in labor supply with respect to S is

$$\frac{\partial L_R}{\partial S} = \frac{\partial L_R}{\partial w_R}\frac{\partial w_R}{\partial S} + \frac{\partial L_R}{\partial I_0}\frac{\partial I_0}{\partial S}.$$

We can substitute the previously derived expressions to get

$$\frac{\partial L_R}{\partial S} = \left(\frac{U_x}{|\Lambda|} - L_R\frac{\Lambda_{11}}{|\Lambda|}\right)\left[(1-\rho)\frac{1}{N}\right] + \left(-\frac{\Lambda_{11}}{|\Lambda|}\right)\rho.$$

Under current policy, in which $\rho = 0$, the effect simplifies to

$$\left.\frac{\partial L_R}{\partial S}\right|_{\rho=0} = \left(\frac{U_x}{|\Lambda|} - L_R\frac{\Lambda_{11}}{|\Lambda|}\right)\left[\frac{1}{N}\right].$$

This is $1/N$ of $\partial L_R/\partial w_R$ in Equation 11. That is, proration has a proportional incentive effect of $1/N$ on labor supplied to the reserve service.

If proration is eliminated, such that $\rho = 1$, then the effect of S on labor supply is

$$\left.\frac{\partial L_R}{\partial S}\right|_{\rho=1} = \left(-\frac{\Lambda_{11}}{|\Lambda|}\right).$$

This is unambiguously negative as long as leisure is a superior good because

$$\frac{\Lambda_{11}}{|\Lambda|} > 0.$$

The policy implication is that paying the full rate of S&I pay instead of prorating it will not increase individual labor supply but will instead decrease it. Therefore, individuals participating in the Selected Reserve who are receiving S&I pays would decrease the intensity of their participation if S&I pays were paid at the full rate rather than being prorated.

A Geometric Explanation of the Extensive Margin

Changing the payment structure of S&I pays could induce RC members (or potential members) who do not show up for training in a given month to do so. This effect is apparently contradictory to the effect described in the previous section, but modeling using RAND's DRM shows that this kind of change would likely increase the number of individuals choosing to participate in the Selected Reserve. On balance, the result of this policy would be to have more individuals participating in the reserve, but they would participate for fewer days each month, on average.

The equations in the previous section do not apply to individuals who are not already performing reserve service, since those equations assume an interior solution: That is, they assume individuals have a strictly positive level of monthly reserve training. For those at a corner (i.e., those with no reserve service), either one of the inequalities in Equations 1 and 2 is strict (i.e., the two sides are not equal) or one of the inequalities in Equation 3 holds with equality.

Instead of using equations, we will illustrate how a change in S&I pay policy can induce new individuals to participate in RC training. The images are based on Shishko and Rostker, 1976, Figure 1.

As in Shishko and Rostker, 1976, we assume that a person's primary job pays wage w_p but that they are constrained to work, at most, L_p units of time per month. Performing reserve service would pay a wage w_R per unit of time, for every unit beyond L_p. When special pays are prorated, this wage includes both basic pay and special pay. The budget set under the proration regime is shown in Figure B.1. We assume, like Shishko and Rostker, that the slope w_R is less than w_p: That is, reserve service pays less per unit of time than the primary job. (If the opposite were true, Shishko and Rostker note that the primary job and secondary jobs could switch.)

Figure B.1 shows a representative indifference curve for a utility-maximizing individual who chooses not to perform reserve service. They choose to work L_0 units of time in their primary job and zero hours of reserve training. This is because w_R (indicated by the slope of the line to the left of L_0) is less than the slope of the indifference curve at the intersection. In more technical terms, the payment for a period of reserve training is less

FIGURE B.1

Example of Nonparticipation Under Special Pay Proration Regime

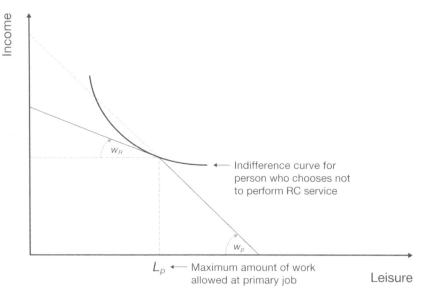

than the marginal rate of substitution between leisure and income. It is not worthwhile to participate in the RC.

In theory, the RC payment rate would need to be increased to the reservation rate (equal to the marginal rate of substitution) before this individual would begin performing reserve training. However, this need not be true if the special pay is offered as a lump sum in return for a minimal amount of reserve service.

Figure B.2 shows how this would happen. Suppose that, instead of being prorated, special pays are offered in the full amount so long as an individual trains a minimum amount of time. The new reserve wage is w_R', which is actually less than w_R because it no longer includes the special pay. Therefore, the new reserve service portion of the budget curve is flatter than before. But now, once the individual trains a minimum amount, the lump sum special pay shifts this budget line vertically upward. This allows the individual to move to a higher indifference curve than before. They now choose to train at least the minimum amount, whereas before they did not train at all.

FIGURE B.2

Example of Induced Participation Under Full-Rate Regime

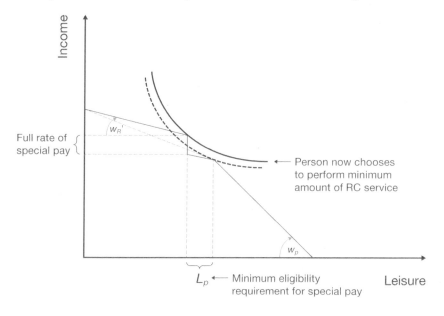

Defining the Analytic Sample

In this appendix, we provide additional details on how we constructed our analytic sample. DMDC provided monthly administrative data for all members of the RC from January 2018 to March 2020. From the Reserve Pay File, we collected information on special pay types and amounts, basic pay for active duty and compensation for inactive duty, and the number of inactive duty drills that were paid. We also collected information on characteristics of service that could affect either a member's eligibility for special pays or whether the pays would be earned during training as opposed to deployment. These characteristics included RCC/TRC code, YAS, the administrative source of pay, the status of the member's pay account, eligibility for full-time active duty benefits, and eligibility for deployment benefits, such as the Combat Zone Tax Exclusion. We excluded data from RC members who did not have any information on service component or RCC/TRC code.

We merged this monthly information with active duty service information from the RC Active Service Transaction File (see DoD Manual 7730.54, 2019), or "Activation File." The Activation File lists active duty stints with an associated statutory authority describing the reason for being called to active duty. Each statutory authority has an associated descriptive project code: For example, AT is the project code corresponding to the statutory authority in 10 U.S.C. § 12301(b).

The Activation File must be transformed into a monthly panel data set that can be merged with the rest of the administrative data. We did this by creating a separate variable for each possible project code. We then counted the number of days served under each project code in each calendar month, using the beginning and end date of each active duty stint.

We wanted to exclude periods when an RC member was performing full-time active duty, either on deployment or in another capacity for which they would count toward AC end strength. However, one caveat regarding the Activation File is that we did not receive data on open active duty stints. In other words, if an RC member was called to active duty under a certain authority and that period of service did not end before March 31, 2020, then we would not receive any information about that service. This is not likely to affect our observation of training stints, since those tend to be short (at worst, someone performing AT in the last week of March 2020 would not have that training recorded in our data). But it is likely to affect deployments. To compensate for these unobserved data, we used an in-house algorithm that marks RC members as deployed according to other observable characteristics. These include assigned location, eligibility for the Combat Zone Tax Exclusion, and having earned IDP.

We used additional indicators that an RC member was serving in a full-time capacity. We excluded months in which an RC member was also listed in the Active Duty Master File, months in which the administrative files said an RC member was eligible for full-time AC health benefits, and months in which the pay file said an RC member was receiving AC pay.

We then excluded RC members according to whether they appeared to be eligible to train for pay. We excluded members with inactive or suspended pay accounts and certain RCC/TRC codes that are not eligible to train with pay (see Figure 2.2).

Figure C.1 shows the sequence of decisions that yielded our final analytic sample.

FIGURE C.1

Decision Steps for Creating the Analytic Sample

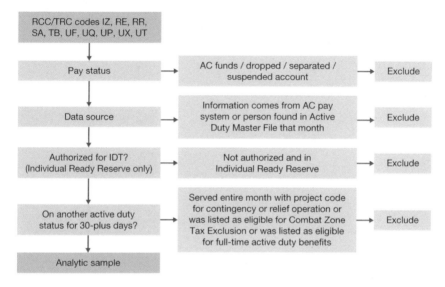

NOTE: This figure shows how DMDC data were cleaned for analysis. The final sample size was 18,571,700 person-months.

Cost Estimates by Reserve Component

In this appendix, we detail the component-specific cost estimates under each of the three methods described in Chapter Three (Tables D.1 through D.6). Baseline costs are the same as those shown in Figures 4.3 and 4.4. The U.S. Coast Guard Reserve is omitted because it is observed to pay only AvIP, with fewer than one RC member earning the pay each month.

TABLE D.1
Cost Estimates for Army Reserve, by Pay and Estimation Method

Pay	Current Cost per Year	Estimated Cost Under Full-Rate Policy, Method 1 (Based on Months of Service)	Estimated Cost Under Full-Rate Policy, Method 2 (Based on Months with Pay)	Estimated Cost Under Full-Rate Policy, Method 3 (Benchmarking Pay Amount Against Periods Served)	Percentage Increase Under Full-Rate Policy, Method 1	Percentage Increase Under Full-Rate Policy, Method 2	Percentage Increase Under Full-Rate Policy, Method 3
AvIP	$1,483,558	$2,515,919	$2,759,879	$4,388,586	70%	86%	196%
Flying duty (crew)	$120,138	$255,793	$226,238	$326,485	113%	88%	172%
Flying duty (noncrew)	$8,740	$21,600	$11,533	$23,832	147%	32%	173%
Diving duty	—	—	—	—	—	—	—
Parachute duty (static)	$242,111	$742,867	$461,667	$740,512	207%	91%	206%
Parachute duty (HALO)	$1,636	$2,031	*	$1,659	24%	*	1%
Demolition duty	$273	$2,133	$600	$1,107	681%	120%	305%
Experimental stress duty	—	—	—	—	—	—	—
Toxic fuels duty	—	—	—	—	—	—	—

SOURCE: DMDC data, January 2018 to March 2020.

NOTE: — indicates that there were no observed payments of that pay by the component during the observation window; * indicates that the cost estimate predicted a decrease in cost and therefore is not reported.

TABLE D.2

Cost Estimates for Army National Guard, by Pay and Estimation Method

Pay	Current Cost per Year	Estimated Cost Under Full-Rate Policy, Method 1 (Based on Months of Service)	Estimated Cost Under Full-Rate Policy, Method 2 (Based on Months with Pay)	Estimated Cost Under Full-Rate Policy, Method 3 (Benchmarking Pay Amount Against Periods Served)	Percentage Increase Under Full-Rate Policy, Method 1	Percentage Increase Under Full-Rate Policy, Method 2	Percentage Increase Under Full-Rate Policy, Method 3
AvIP	$10,054,324	$17,426,297	$20,384,273	$33,504,462	73%	103%	233%
Flying duty (crew)	$1,680,504	$4,227,289	$4,332,744	$6,534,587	151%	157%	289%
Flying duty (noncrew)	$163,815	$467,467	$387,800	$624,564	185%	137%	281%
Diving duty	$36,996	*	*	$79,288	*	*	114%
Parachute duty (static)	$2,581,022	$3,468,333	$4,573,400	$7,829,567	34%	77%	203%
Parachute duty (HALO)	$48,811	$117,100	$97,700	$102,119	140%	100%	109%
Demolition duty	$299,976	$548,467	$484,933	$972,487	83%	62%	224%
Experimental stress duty	$162	$1,600	$333	$278	888%	105%	71%
Toxic fuels duty	—	—	—	—	—	—	—

SOURCE: DMDC data, January 2018 to March 2020.

NOTE: — indicates that there were no observed payments of that pay by the component during the observation window; * indicates that the cost estimate predicted a decrease in cost and therefore is not reported.

TABLE D.3
Cost Estimates for Air Force Reserve, by Pay and Estimation Method

Pay	Current Cost per Year	Estimated Cost Under Full-Rate Policy, Method 1 (Based on Months of Service)	Estimated Cost Under Full-Rate Policy, Method 2 (Based on Months with Pay)	Estimated Cost Under Full-Rate Policy, Method 3 (Benchmarking Pay Amount Against Periods Served)	Percentage Increase Under Full-Rate Policy, Method 1	Percentage Increase Under Full-Rate Policy, Method 2	Percentage Increase Under Full-Rate Policy, Method 3
AvIP	$11,482,126	$25,290,933	$28,399,822	$31,712,931	120%	147%	176%
Flying duty (crew)	$492,464	$1,482,529	$1,275,964	$1,667,651	201%	159%	239%
Flying duty (noncrew)	$70,023	$222,400	$164,467	$253,972	218%	135%	263%
Diving duty	$4,870	$7,200	*	$8,110	48%	*	67%
Parachute duty (static)	$6,115	$12,600	$6,800	$14,015	106%	11%	129%
Parachute duty (HALO)	$12,596	$34,400	*	$17,238	173%	*	37%
Demolition duty	$77,016	$162,533	$167,467	$262,708	111%	119%	241%
Experimental stress duty	$176,508	*	*	$302,117	*	*	71%
Toxic fuels duty	—	—	—	—	—	—	—

SOURCE: DMDC data, January 2018 to March 2020.

NOTE: — indicates that there were no observed payments of that pay by the component during the observation window; * indicates that the cost estimate predicted a decrease in cost and therefore is not reported.

TABLE D.4
Cost Estimates for Air National Guard, by Pay and Estimation Method

Pay	Current Cost per Year	Estimated Cost Under Full-Rate Policy, Method 1 (Based on Months of Service)	Estimated Cost Under Full-Rate Policy, Method 2 (Based on Months with Pay)	Estimated Cost Under Full-Rate Policy, Method 3 (Benchmarking Pay Amount Against Periods Served)	Percentage Increase Under Full-Rate Policy, Method 1	Percentage Increase Under Full-Rate Policy, Method 2	Percentage Increase Under Full-Rate Policy, Method 3
AvIP	$11,625,385	$22,110,822	$25,694,933	$28,657,748	90%	121%	147%
Flying duty (crew)	$263,380	$925,607	$676,780	$907,018	251%	157%	244%
Flying duty (noncrew)	$65,992	$233,533	$189,867	$283,166	254%	188%	329%
Diving duty	$4,747	$7,468	*	$6,360	57%	*	34%
Parachute duty (static)	$15,569	$24,533	*	$28,358	58%	*	82%
Parachute duty (HALO)	$4,254	$11,700	*	$7,186	175%	*	69%
Demolition duty	$60,947	$142,000	$113,133	$170,048	133%	86%	179%
Experimental stress duty	$283,387	*	*	$443,214	*	*	56%
Toxic fuels duty	$1,789	$7,400	*	$8,660	314%	*	384%

SOURCE: DMDC data, January 2018 to March 2020.

NOTE: – indicates that there were no observed payments of that pay by the component during the observation window; * indicates that the cost estimate predicted a decrease in cost and therefore is not reported.

TABLE D.5
Cost Estimates for Marine Corps Reserve, by Pay and Estimation Method

Pay	Current Cost per Year	Estimated Cost Under Full-Rate Policy, Method 1 (Based on Months of Service)	Estimated Cost Under Full-Rate Policy, Method 2 (Based on Months with Pay)	Estimated Cost Under Full-Rate Policy, Method 3 (Benchmarking Pay Amount Against Periods Served)	Percentage Increase Under Full-Rate Policy, Method 1	Percentage Increase Under Full-Rate Policy, Method 2	Percentage Increase Under Full-Rate Policy, Method 3
AvIP	$1,027,127	$3,088,993	$3,088,113	$3,512,908	201%	201%	242%
Flying duty (crew)	$11,502	$67,378	$22,511	$47,634	486%	96%	314%
Flying duty (noncrew)	$298	$2,667	$667	$1,453	795%	124%	388%
Diving duty	$3,444	$12,127	*	$11,107	252%	*	223%
Parachute duty (static)	$8,267	$86,133	$18,933	$33,564	941%	129%	306%
Parachute duty (HALO)	$2,693	$14,700	$2,700	$8,985	446%	<1%	234%
Demolition duty	—	—	—	—	—	—	—
Experimental stress duty	—	—	—	—	—	—	—
Toxic fuels duty	—	—	—	—	—	—	—

SOURCE: DMDC data, January 2018 to March 2020.

NOTE: — indicates that there were no observed payments of that pay by the component during the observation window; * indicates that the cost estimate predicted a decrease in cost and therefore is not reported.

TABLE D.6

Cost Estimates for Navy Reserve, by Pay and Estimation Method

Pay	Current Cost per Year	Estimated Cost Under Full-Rate Policy, Method 1 (Based on Months of Service)	Estimated Cost Under Full-Rate Policy, Method 2 (Based on Months with Pay)	Estimated Cost Under Full-Rate Policy, Method 3 (Benchmarking Pay Amount Against Periods Served)	Percentage Increase Under Full-Rate Policy, Method 1	Percentage Increase Under Full-Rate Policy, Method 2	Percentage Increase Under Full-Rate Policy, Method 3
AvIP	$3,157,573	$7,475,517	$7,710,067	$10,401,616	137%	144%	229%
Flying duty (crew)	$31,643	$89,236	$69,827	$113,671	182%	121%	259%
Flying duty (noncrew)	$5,453	$21,200	$17,000	$21,448	289%	212%	293%
Diving duty	$24,608	$102,136	$34,529	$82,972	315%	40%	237%
Parachute duty (static)	$2,962	$25,467	$6,200	$10,179	760%	109%	244%
Parachute duty (HALO)	$3,235	$3,400	*	$7,438	5%	*	130%
Demolition duty	$7,934	$41,067	$12,267	$22,899	418%	55%	189%
Experimental stress duty	—	—	—	—	—	—	—
Toxic fuels duty	—	—	—	—	—	—	—

SOURCE: DMDC data, January 2018 to March 2020.

NOTE: — indicates that there were no observed payments of that pay by the component during the observation window; * indicates that the cost estimate predicted a decrease in cost and therefore is not reported.

Data Quality Metrics

In this appendix, we provide additional comparisons between active duty service and special pay data. Tables E.1 and E.2 illustrate patterns in the data that make it difficult to calculate precisely how many months an RC member performed hazardous duties. These patterns, in addition to those discussed in Chapters Three and Four, help explain why our various cost estimation methods yield different answers. They also help explain why some of our estimates were lower than actual costs (and therefore were not reported in the main text or in Appendix D).

Periods Paid Often Outnumber Periods Served

To check how pay lined up with service, we aggregated the number of periods compensated by each special pay for each RC member during 2018 and 2019. We also aggregated the number of active duty periods served and the number of IDT drills paid to estimate total periods served. Table E.1 shows the average number of special pay periods paid versus periods served for those who earned each pay at some point during a calendar year.

Table E.1 shows that, according to the data, many if not most RC members who earn special pays are compensated for more periods of hazardous duty than they are observed to have served. It is not possible to determine the source of this apparent discrepancy. One explanation is that certain types of active duty service are not reported in the RC Active Service Transaction File.

Annual Training Appears to Be Underreported

To test whether some active service could be unreported, we counted how many drilling reservists have AT reported in the RC Active Service Transac-

TABLE E.1
Annual Periods Served Versus Periods Compensated by Special Pay for RC Members Who Earned Pay at Least Once in 2018 or 2019

	Flying Duty (Crew)	Flying Duty (Noncrew)	Diving Duty	Parachute Duty (Static)	Parachute Duty (HALO)	Flight Deck Duty	Demolition Duty	Experimental Stress Duty	Toxic Fuels Duty	AvIP
Average periods served in calendar year	68	61	58	54	78	74	53	63	74	83
Average periods compensated by special pay in calendar year	92	73	71	101	53	23	75	343	67	132
Percentage of payees with more periods paid than served	67%	53%	43%	70%	32%	0%	55%	82%	33%	66%

SOURCE: DMDC data.

NOTE: Periods served equals total active duty days listed in the RC Active Service Transaction File plus IDT drills paid, as listed in the Reserve Pay File. Periods compensated is calculated by dividing the special pay amount by the per-period rate.

tion File at any time during 2018 and 2019. Table E.2 shows the percentage of RC members in each component who have any AT reported. The Coast Guard Reserve is omitted because it did not report special pays (except for rare instances of AvIP).

Table E.2 shows that only a minority of drilling reservists have any AT reported, ranging from a high of 39.3 percent in the Air Force Reserve to just 1.6 percent in the Navy Reserve. But when AT is reported, the length of time is as expected: The median length of AT is approximately equal to the required 14 days, except for the Navy Reserve.

To check whether AT might be reported as another duty status, Table E.2 also shows the percentage of drilling reservists who had any active duty service reported, including all forms of ADT as well as active duty other than training or a brief (less than one month) mobilization. Still, only a minority of members had any such service reported. Therefore, even if AT is being reported under the statutory authority for another active duty status, it appears to be underrepresented in the data we received.

The reasons for these patterns are unclear. It could be that AT stints are opened but not closed in the RC Active Service Transaction File: That is, they have a start date, but an end date is never entered. In that case, we would not receive that information because we receive only the closed active duty segments. It is also possible that AT is not reported at all, either because personnel systems are not designed to record this information in a systematic way that feeds into the transaction file or because it is administratively burdensome to record every AT segment.

Assuming that we are not able to observe every AT segment but that we still observe all payments of S&I pays, our cost estimates could be affected in two ways. First, our estimate using only service data (Method 1) could be low. If AT is performed in months in which we otherwise observe no service, then there are more months of service than are included in our count, meaning we could underestimate the number of months of S&I pay eligibility.

Second, our estimate that benchmarks pay to service (Method 3) could be affected. If 14 days of AT were missing from most RC members' transaction record, our calculation of the average periods of service per month would be too low. This means that Method 3 would estimate that pay disbursements are compensating for more months than they actually are, making our counterfactual estimates too high.

TABLE E.2

Percentage of Drilling Reservists with Any Active Duty Service Reported in Transaction File During a Full Calendar Year, by Duty Status and Component

	Army Reserve	Army National Guard	Air Force Reserve	Air National Guard	Marine Corps Reserve	Navy Reserve
Percentage with any AT reported	26.9%	<1%	39.3%	33.0%	30.7%	1.6%
Median AT days, if more than zero	18	14	15	13	15	90
Percentage with any active duty status reported	35.6%	8.4%	43.4%	43.0%	32.7%	19.7%
Median active duty days, if more than zero	23	41	22	24	15	14

SOURCE: DMDC data.

NOTE: Drilling reservists are RC members with RCC/TRC code "SA," who are typically expected to serve 14 days of AT each year.

References

Asch, Beth J., James V. Marrone, and Michael G. Mattock, *An Examination of the Methodology for Awarding Imminent Danger Pay and Hostile Fire Pay*, Santa Monica, Calif.: RAND Corporation, RR-3231-OSD, 2019. As of August 4, 2021:
https://www.rand.org/pubs/research_reports/RR3231.html

Defense Finance and Accounting Service, "Military Pay Tables & Information," webpage, undated. As of October 11, 2020:
https://www.dfas.mil/militarymembers/payentitlements/Pay-Tables/

Defense Manpower Data Center, "DoD Personnel, Workforce Reports & Publications," webpage, undated. As of August 2, 2021:
https://dwp.dmdc.osd.mil/dwp/app/dod-data-reports/workforce-reports

Department of Defense Instruction 1215.06, *Uniform Reserve, Training, and Retirement Categories for the Reserve Components*, Washington, D.C.: U.S. Department of Defense, May 19, 2015. As of October 9, 2020:
https://www.esd.whs.mil/Portals/54/Documents/DD/issuances/dodi/121506p.pdf

Department of Defense Instruction 1340.09, *Hazard Pay (HzP) Program*, Washington, D.C.: U.S. Department of Defense, January 26, 2018. As of October 9, 2020:
https://www.esd.whs.mil/Portals/54/Documents/DD/issuances/dodi/134009p.PDF

Department of Defense Instruction 7730.67, *Aviation Incentive Pays and Bonus Program*, Washington, D.C.: U.S. Department of Defense, October 20, 2016. As of October 9, 2020:
https://www.esd.whs.mil/Portals/54/Documents/DD/issuances/dodi/773067_dodi_2016.pdf

Department of Defense Instruction 7770.02, *Uniformed Services Pay File*, Washington, D.C.: U.S. Department of Defense, June 10, 2019. As of July 29, 2021:
https://www.esd.whs.mil/Portals/54/Documents/DD/issuances/dodi/777002p.pdf

Department of Defense Manual 7730.54, Volume 1, *Reserve Components Common Personnel Data System (RCCPDS): Reporting Procedures*, Washington, D.C.: U.S. Department of Defense, January 28, 2019. As of July 29, 2021:
https://www.esd.whs.mil/Portals/54/Documents/DD/issuances/dodm/773054m_vol1.pdf

DMDC—*See* Defense Manpower Data Center.

DoD—*See* U.S. Department of Defense.

DoDI—*See* Department of Defense Instruction.

Gilroy, Curtis L., David K. Horne, and D. Alton Smith, eds., *Military Compensation and Personnel Retention: Models and Evidence*, Alexandria, Va.: U.S. Army Research Institute for the Behavioral and Social Sciences, February 1991.

Hogan, Paul F., Patrick C. Mackin, Captain Louis M. Farrell, and Captain George T. Elliott, "Special and Incentive Pays for the Reserve Components," in U.S. Department of Defense, *Report of the Ninth Quadrennial Review of Military Compensation*: Vol. III, *Creating Differentials in Military Pay: Special and Incentive Pays*, Washington, D.C., March 2002, pp. 135–146.

Kapp, Lawrence, and Barbara Salazar Torreon, *Reserve Component Personnel Issues: Questions and Answers*, Washington, D.C.: Congressional Research Service, RL30802, updated June 15, 2020. As of July 26, 2021:
https://crsreports.congress.gov/product/details?prodcode=RL30802

Kocher, Kathryn, and George Thomas, *Gender Differences in the Retention of Enlisted Army Reservists*, Monterey, Calif.: Naval Postgraduate School, November 1990.

Levinsky, David, "A New Bill Would Raise Hazard Pay for Reservists and Guardsmen to Equal Active Duty's," *Task & Purpose*, February 21, 2020. As of April 20, 2021:
https://taskandpurpose.com/news/hazard-pay-national-guard-reservists/

Mattock, Michael G., James Hosek, Beth J. Asch, and Rita Karam, *Retaining U.S. Air Force Pilots When the Civilian Demand for Pilots Is Growing*, Santa Monica, Calif.: RAND Corporation, RR-1455-AF, 2016. As of August 2, 2021:
https://www.rand.org/pubs/research_reports/RR1455.html

Mehay, Stephen L., "Reserve Participation Versus Moonlighting: Are They the Same?" *Defence and Peace Economics*, Vol. 2, No. 4, 1991, pp. 325–337.

Office of the Under Secretary of Defense (Comptroller), *Financial Management Regulation*, Vol. 7A: *Military Pay Policy—Active Duty and Reserve Pay*, Washington, D.C.: U.S. Department of Defense, DoD 7000.14-R, May 2020. As of December 22, 2020:
https://comptroller.defense.gov/Portals/45/documents/fmr/Volume_07a.pdf

———, *Military Personnel Programs (M-1): Department of Defense Budget Fiscal Year 2022*, Washington, D.C.: U.S. Department of Defense, May 2021. As of August 2, 2021:
https://comptroller.defense.gov/Portals/45/Documents/defbudget/FY2022/FY2022_m1.pdf

Public Law 116-92, National Defense Authorization Act for Fiscal Year 2020, December 20, 2019.

Public Law 116-283, William M. (Mac) Thornberry National Defense Authorization Act for Fiscal Year 2021, January 1, 2021.

Shishko, Robert, and Bernard Rostker, "The Economics of Multiple Job Holding," *American Economic Review*, Vol. 66, No. 3, June 1976, pp. 298–308.

Titus, Cory, "Reservists Deserve the Same Pay for the Same Risks," Military Officers Association of America, March 16, 2020. As of April 20, 2021: https://www.moaa.org/content/publications-and-media/ news-articles/2020-news-articles/advocacy/reservists-deserve-the-same-pa y-for-the-same-risks/

U.S. Code, Title 10, Section 10102, Purpose of Reserve Components, January 3, 2012.

U.S. Code, Title 10, Section 10147, Ready Reserve: Training Requirements, January 3, 2012.

U.S. Code, Title 10, Section 12301, Reserve Components Generally, January 3, 2007.

U.S. Code, Title 10, Section 12302, Ready Reserve, January 3, 2012.

U.S. Code, Title 10, Section 12304, Selected Reserve and Certain Individual Ready Reserve Members; Order to Active Duty Other Than During War or National Emergency, January 3, 2012.

U.S. Code, Title 10, Section 12319, Ready Reserve: Muster Duty, January 3, 2012.

U.S. Code, Title 10, Section 12503, Ready Reserve: Funeral Honors Duty, January 3, 2007.

U.S. Code, Title 37, Section 204, Entitlement, January 3, 2012.

U.S. Code, Title 37, Section 206, Reserves; Members of National Guard: Inactive-Duty Training, January 7, 2011.

U.S. Code, Title 37, Section 301, Incentive Pay: Hazardous Duty, January 7, 2011.

U.S. Code, Title 37, Section 304, Special Pay: Diving Duty, January 3, 2016.

U.S. Code, Title 37, Section 353, Skill Incentive Pay or Proficiency Bonus, January 3, 2012.

U.S. Code, Title 37, Section 334, Special Aviation Incentive Pay and Bonus Authorities for Officers, January 3, 2012.

U.S. Code, Title 37, Section 351, Hazardous Duty Pay, January 3, 2012.

U.S. Code, Title 37, Section 433, Allowance for Muster Duty, January 7, 2011.

U.S. Code, Title 37, Section 495, Funeral Honors Duty: Allowance, January 16, 2014.

U.S. Code, Title 37, Section 502, Absences Due to Sickness, Wounds, and Certain Other Causes, January 7, 2011.

U.S. Department of Defense, *Report of the Sixth Quadrennial Review of Military Compensation,* Volume I, Washington, D.C., August 1988.

———, *Report of the Ninth Quadrennial Review of Military Compensation,* Volume I, Washington, D.C., March 2002.

———, *Report of the Eleventh Quadrennial Review of Military Compensation, Main Report,* Washington, D.C., June 2012.

———, "Military Personnel Guidance (Supplement 1) for DoD and OSD Component Heads and Military Commanders in Responding to Coronavirus Disease 2019 (COVID-19)," April 9, 2020. As of June 8, 2021: https://www.whs.mil/Portals/75/Coronavirus/DOD%20Mil%20Pers%20 Guidance%20Supp%201.pdf